Mars Wars
The Rise and Fall of the
Space Exploration Initiative

Thor Hogan

Mars Wars
The Rise and Fall of the Space Exploration Initiative

Thor Hogan

The NASA History Series

National Aeronautics and Space Administration
NASA History Division
Office of External Relations
Washington, DC
August 2007
NASA SP-2007-4410

Library of Congress Cataloging-in-Publication Data

Hogan, Thor.
  Mars wars : the rise and fall of the Space Exploration Initiative / Thor Hogan.
    p. cm. -- (The NASA history series) (NASA SP-2007-4410)
    Includes bibliographical references and index.
    1. Space Exploration Initiative (U.S.) 2. Space flight to Mars--Planning--History--20th century. 3. United States. National Aeronautics and Space Administration--Management--History--20th century.
  4. Astronautics and state--United States--History--20th century.
  5. United States--Politics and government--1989-1993.
  6. Outerspace--Exploration--United States--History--20th century.
  7. Organizational change--United States--History--20th century. I. Title.
  TL789.8.U6S62 2007
  629.45'530973--dc22

                                              2007008987

Table of Contents

Acknowledgements. iii
Chapter 1: Introduction 1
   The Policy Stream and Punctuated Equilibrium Models. 2
   Why Mars?. 5
      Canals on Mars. 7
      Mars in Popular Culture 9
      Mariner and Viking 11
Chapter 2: The Origins of SEI 15
   Early Mission Planning 16
   Post-Apollo Planning 21
   Case for Mars. 25
   National Commission on Space 27
   The Ride Report 30
   President Reagan and NASA's Office of Exploration. 32
Chapter 3: Bush, Quayle, and SEI. 37
   Bush-Quayle 1988 39
   Reagan-Bush Transition 44
   The Problem Stream: Providing Direction to a Directionless Agency. 47
   The Policy Stream: The Ad Hoc Working Group 57
   The Political Stream: Briefing Key Actors 64
   Joining the Streams: Human Exploration of Mars Reaches the Gov. Agenda . 68
Chapter 4: The 90-Day Study. 77
   Waiting for NASA 82
   SEI Takes Shape. 85
   Mars Wars 92
Chapter 5: The Battle to Save SEI. 107
   Presidential Decisions. 108
   SEI Hits the Road 114
   Losing Faith in NASA 120
   The Augustine and Synthesis Group Reports 125
   SEI Fades Away 130
Chapter 6: SEI, Policy Streams, and Punctuated Equilibrium 137
   Policy Streams, SEI, and the Space Policy Community. 138
   Punctuated Equilibrium, Space Policy, and SEI 143
Chapter 7: The Lessons of SEI 159
NASA History Series 167
Index 178

Acknowledgements

A great many people provided me with assistance during the preparation of this book. I am particularly indebted to John Logsdon and Roger Launius for their invaluable insights regarding the history of the American space program, the role of various actors within the space policy community, and the context in which Mars exploration should be viewed. I also owe a great deal to Steve Balla and Jeff Henig, who helped me employ the political science and public policy theories that are used in this book. I would also like to thank Ray Williamson and Joe Cordes for their support and helpful comments. I am also very grateful for all the hard work contributed by Steve Dick and Steve Garber getting this book ready for publication. I would like to thank Heidi Pongratz at Maryland Composition, Angela Lane and Danny Nowlin at Stennis Space Center, and David Dixon at NASA Headquarters for handling the copyediting, layout, and printing of the book. Finally, I would like to thank the dedicated archivists at both the NASA History Division in Washington, D.C. and the George Bush Presidential Library at Texas A&M – they were all crucial to the successful completion of this project. In particular, I would like to thank Colin Fries and Jane Odom at the NASA History Division and Debbie Carter, Bob Holzweiss, John Laster, Laura Spencer, and Melissa Walker at the George Bush Presidential Library.

This book is dedicated to Joe Hogan, for teaching me how to dream big dreams; to Ron Beck, for believing in my potential when few others did; and to Kate Kuvalanka, for inspiring me on a daily basis

1

Introduction

"Why Mars? Because it is humanity's destiny to strive, to seek, to find. And because it is America's destiny to lead."

President George Bush, 20 July 1989[1]

Sitting on the steps of the Smithsonian Institution's National Air and Space Museum on 20 July 2004, it was difficult to imagine that fifteen years earlier at this place the fiercest domestic political conflict of the first space age commenced. Bright-eyed children poured off tour buses and hurried to examine the museum's wonders, teachers and parents close behind with digital cameras ready. They lined up to touch a four billion year-old lunar rock, clamored around the Apollo 11 Command Module *Columbia*, and gazed up at an ungainly Lunar Module— monuments to past American triumphs in space. It was on a similarly hot and muggy Washington morning that President George Bush had used this location to announce a renewed commitment to human exploration beyond Earth orbit. Even before this declaration, however, the winds of war had been swirling in the nation's capital.

On the 20th anniversary of the first human landing on the Moon, President Bush had stood atop these very steps and proposed a long-range exploration plan that included the successful construction of an orbital space station, a permanent return to the Moon, and a human mission to Mars—this enterprise became known as the Space Exploration Initiative (SEI). The president charged the newly reestablished National Space Council with providing concrete alternatives for meeting these objectives. To provide overall focus for the new initiative, Bush later set a 30-year goal for a crewed landing on Mars. If met, humans would be walking on the red

---

[1] *Public Papers of the Presidents of the United States*, 20 July 1989, *Remarks on the 20th Anniversary of the Apollo 11 Moon Landing* [http://bushlibrary.tamu.edu/papers/] (accessed 18 May 2002).

planet by 2019, which would be the 50th anniversary of the Apollo 11 lunar landing. Within a few short years after President Bush's Kennedyesque announcement, however, the initiative had faded into history—the victim of a flawed policy process and a political war fought on several different fronts. The failure of SEI, combined with problems ranging from the Hubble Space Telescope's flawed mirror to space shuttle fuel leaks to space station budget problems, badly damaged the National Aeronautics and Space Administration's (NASA) image and prompted dramatic changes in the American space program.

The rise of SEI and its eventual demise represents one of the landmark episodes in the history of the American space program—ranking with the creation of NASA, the decision to go to the Moon, the post-Apollo planning process, and the space station decision. The story of this failed initiative is one shaped by key protagonists and critical battles. It is a tale of organizational, cultural, and personal confrontation. Organizational skirmishes involved the Space Council versus NASA, the White House versus congressional appropriators, and the Johnson Space Center versus the rest of the space agency—all seeking control of the national space policy process. Cultural struggles pitted the increasingly conservative engineering ethos of NASA against the "faster, better, cheaper" philosophy of a Space Council looking for innovative solutions to technical problems. Personality clashes matched Vice President Dan Quayle and Space Council Executive Secretary Mark Albrecht against NASA Administrator Dick Truly and Johnson Space Center Director Aaron Cohen. In the final analysis, the demise of SEI was a classic example of a defective decision-making process—one that lacked adequate high-level policy guidance, failed to address critical fiscal constraints, developed inadequate programmatic alternatives, and garnered no congressional support. Some space policy experts have argued that SEI was doomed to fail, due primarily to the immense budgetary pressures facing the nation during the early 1990s.[2] This book will argue, however, that the failure of the initiative was not predetermined; instead, it was the result of a deeply flawed policy process that failed to develop (or even consider) policy options that may have been politically acceptable given the existing political environment.

The Policy Stream and Punctuated Equilibrium Models

To make the case that the failure of SEI was not inevitable, this study employs two theoretical models to guide the narrative analysis of how the initiative reached

---

[2] Dwayne A. Day, "Doomed to Fail: The Birth and Death of the Space Exploration Initiative," *Spaceflight* (March 1995), pp. 79-83; John Pike, "But what is the True Rationale for Human Spaceflight?," *Space Policy* (August 1994), pp. 217-222.

Chapter 1: Introduction

the government agenda and what factors led to its ultimate demise. John Kingdon's *Policy Streams Model* describes how problems come to the attention of policy makers, how agendas are set, how policy alternatives are generated, and why policy windows open.[3] This theory will be utilized to develop the story of SEI's rise and fall, and will more specifically be used to assess who the important actors are within the space policy community. Frank Baumgartner and Bryan Jones's *Punctuated Equilibrium Model* depicts the policy process as comprising long periods of stability, which are interrupted by predictable periods of instability that lead to major policy changes.[4] This model will be utilized to provide a better understanding of the larger trends that led to SEI's promotion to the government agenda and may explain its eventual downfall.[5] These two models contributed a number of descriptive statistics that were used to develop a collection of lessons learned from the SEI experience.

In 1972, Michael Cohen, James March, and Johan Olsen introduced *Garbage Can Theory* in an article describing what they called "organized anarchies." The authors emphasized the chaotic character of organizations as loose collections of ideas as opposed to rational, coherent structures. They found that each, organized anarchy was composed of four separate process streams: problems, solutions, participants, and choice opportunities. They concluded that organizations are "a collection of choices looking for problems, issues and feelings looking for decision situations in which they might be aired, solutions looking for issues to which they might be the answer, and decision makers looking for work." Finally, a choice opportunity was:

> …a garbage can into which various kinds of problems and solutions are dumped by participants as they are generated. The mix of garbage in a single can depends on the mix of cans available, on the labels attached to the alternative cans, on what garbage is currently being produced, and on the speed with which garbage is collected and removed from the scene.

Therefore, the three found that policy outcomes are the result of the garbage available and the process chosen to sift through that garbage.[6]

---

[3] John W. Kingdon, *Agendas, Alternatives, and Public Policies* (New York: HarperCollins College Publishers, 1995).

[4] Frank R. Baumgartner and Bryan D. Jones, *Agendas and Instability in American Politics* (Chicago: University of Chicago Press, 1993), pp. 3-24.

[5] Wayne Parsons, *Public Policy: An Introduction to the Theory and Practice of Policy Analysis* (Brookfield, Vermont: Edward Elgar Publishing, 1995), pp. 193-207.

[6] Michael Cohen, James March, and Johan Olsen, "A Garbage Can Model of Organizational Choice," *Administrative Science Quarterly* (March 1972), pp. 1-25; Parsons, *Public Policy*, pp. 192-193.

Mars Wars

In his classic tome *Agendas, Alternatives and Public Policies*, John Kingdon applies the garbage can model to develop a framework for understanding the policy process within the federal government. He found that there were three major process streams in federal policy making: problem recognition; the formation and refinement of policy proposals; and politics. Kingdon concludes that these three process streams operate largely independent from one another. Within the first stream, various problems come to capture the attention of people in and around government. Within the second stream, a policy community of specialists concentrates on generating policy alternatives that may offer a solution to a given problem. Within the third stream, phenomena such as changes in administration, shifts in partisan or ideological distributions in Congress, and focusing events impact the selection of different policy alternatives. Kingdon argues that the key to gaining successful policy outcomes within this "organized anarchy" is to seize upon policy windows that offer an opportunity for pushing one's proposals onto the policy agenda. Taking advantage of these policy windows requires that a policy entrepreneur expend the political capital necessary to join the three process streams at the appropriate time.[7] Kingdon's model provided a useful structure for assessing the role of the policy community in placing SEI on the government agenda and formulating alternatives to solve a perceived problem—a lack of strategic direction within the American space program. Furthermore, it provided benchmarks that were used to evaluate the flawed policy making process for the initiative. In particular, it provided an analytic tool for understanding why Vice President Quayle and Mark Albrecht were not able to successfully join the three process streams when a policy window opened for human exploration beyond Earth orbit.

In *Agendas and Instability in American Politics*, Frank Baumgartner and Bryan Jones introduce a punctuated equilibrium model of policy change in American politics, based on the emergence and recession of policy issues on the government agenda. This theory suggests that the policy process has long periods of equilibrium, which are periodically disrupted by some instability that results in dramatic policy change. Baumgartner and Jones describe "a political system that displays considerable stability with regard to the manner in which it processes issues, but the stability is punctuated with periods of volatile change." Within this system, they contend, the mass public is limited in its ability to process information and remain focused on any one issue. As a result, policy subsystems are created so that scores of agenda items can be processed simultaneously. Only in times of unique crisis and instability

---

[7] John W. Kingdon, *Agendas, Alternatives, and Public Policies* (New York: HarperCollins College Publishers, 1995), pp. 86-89.

do issues rise to the top of the government agenda to be dealt with independently. At a fundamental level, the punctuated equilibrium model seeks to explain why the policy process is largely incremental and conservative, but is also subject to periods of radical change.[8]

Baumgartner and Jones argue that to understand the complexities of the policy making process, one must study specific policy problems over extended periods of time. To comprehend the policy dynamics of an issue, one must develop indicators that explain how the issue is understood. They introduce a new approach to policy research that attempts to meld the policy typology literature and the agenda status literature—the former based on cross-sectional comparisons of multiple public policy issues, the latter focused on longitudinal studies of a single issue over time. The new approach concentrates on the long-term trends related to interest in, and discussion of, important policy questions. In particular, they are interested in two related concepts, whether an issue is on the agenda of a given institution (venue access) and whether the tone of activity within that institution is positive or negative (policy image).[9] The two utilize an eclectic group of measures to gauge venue access and policy image. Baumgartner and Jones's model provided a useful method for understanding where SEI fits within the history of the American space program. More importantly, it provided a means to evaluate whether long-term space policy trends predetermined the initiative's failed outcome.

Why Mars?

Any discussion of human exploration of Mars must begin with a description of the reasons why this planetary destination has continually reemerged during the post-Apollo period as the "next logical step" for the American space program. Understanding the deep-rooted human interest in Mars provides some insight into the space program's recurring focus on it as an objective for both robotic and human missions. Crewed Mars exploration has been seriously considered three times during the past 35 years, but our fascination with the red planet began a great deal earlier. For thousands of years, the human race has been drawn to Mars—our celestial neighbor fuels the imagination unlike any other planet in the solar system. Ancient humans examined the red planet as they attempted to unlock the mystery of the heavens. To primitive humans, the fourth planet from the sun was nothing more then a reddish point of light dancing across the night sky. Early civilizations gave

---

[8] Frank R. Baumgartner and Bryan D. Jones, *Agendas and Instability in American Politics* (Chicago: University of Chicago Press, 1993), pp. 3-24.

[9] Ibid., pp. 39-55.

it many names: the Egyptians called it Har decher (the Red One), the Babylonians named it Nergal (the Star of Death), the Greeks designated it Ares and the Romans called it Mars (both representing the God of War). While the early Babylonians made extensive astronomical observations, it was the Greeks that first categorized Ares as one of five wandering "planets" among the fixed stars (the others being Mercury, Venus, Jupiter, and Saturn). Greek astronomers observed that Ares did not always move from east to west, but sometimes moved in the opposite direction. Due to the existing belief that the Earth was the center of the universe, this astronomical oddity would baffle sky watchers for centuries to come. By 250 B.C., Aristarchus of Samos had developed a complete heliocentric system that viewed Earth as an ordinary planet circling the sun once every year. This theory held the key to understanding the unusual movements of Ares. Later Greek and Roman astronomers did not follow Aristarchus's lead, however, choosing to hold onto the geocentric system. Claudius Ptolemy made the greatest elaboration of this system during the second century A.D.—his geocentric model remained the predominant astronomical theory for more than a millennium.[10]

Seventeen hundred years after Aristarchus first developed it, a Polish canon named Nicolaus Copernicus reintroduced the heliocentric model. Like Aristarchus, however, Copernicus could not exactly predict the motions of the planets using simple circular orbits. As a result, his contemporaries largely ignored his theories. While Copernicus had been primarily a theoretician, it would take two dedicated observational astronomers to discover the true movements of the planets—their names were Tycho Brahe and Johannes Kepler. Starting in 1576, Tycho spent 20 years studying the motions of the stars and planets, including Mars. In 1600, Kepler joined him and began examining the apparent retrograde motion of Mars. When Tycho died the next year, Kepler was appointed to succeed him as Imperial Mathematician to the Holy Roman Emperor (although he was Lutheran).[11]

Using Tycho's scrupulous observations, Kepler went to work trying to explain Mars' apparent backward motion. Kepler argued that the planets revolved around the sun, but at different distances and therefore different speeds. While Earth orbited the sun in 365 days, it took Mars 687 days. Thus, the retrograde movement of Mars could be explained because the Earth was overtaking the slower-moving Mars. To an observer on Earth, it would appear that Mars was slowing down and then reversing course. Kepler proved, however, that this was simply an illusion. In 1609, Kepler published *On the Motion of Mars*, which expounded his first two laws of planetary

---

[10] William Sheehan, *The Planet Mars: A History of Observation and Discover* (Tucson: The University of Arizona Press, 1996), pp. 1-8; John Noble Wilford, *Mars Beckons: The Mysteries, the Challenges, the Expectations of Our Next Great Adventure in Space* (New York: Vintage Books, 1990), pp. 3-17.

[11] Ibid.

motion—stating that planetary orbits about the Sun were elliptical (as opposed to circular as Aristarchus and Copernicus had assumed) and that a planet's speed increases as it approaches the sun and decreases proportionally as it moves farther away. As a result of Tycho and Kepler's observations and theories, the heliocentric system finally overcame Ptolemy's geocentric model.[12]

In 1609, the same year that Kepler published *On the Motion of Mars*, Galileo Galilei made the first celestial observations with a telescope. The next year, after making observations of the Moon, Jupiter, and Venus, Galileo turned his telescope toward Mars. Due to the use of a relatively crude instrument, Galileo's observations of Mars where not particularly informative—other than to suggest that the planet was not a perfect sphere. In 1659, Dutch astronomer Christian Huygens, using a considerably more advanced telescope, was able to detect the first surface feature on Mars. The dark triangular area that he observed over a period of months, which is today called Syrtis Major, allowed him to conclude that Mars rotated on its axis like the Earth. Seven years later, in 1666, Italian astronomer Giovanni Cassini began a series of observations and discovered the planet's white polar caps.[13]

In 1783, astronomer William Herschel, who two years earlier had discovered the planet Uranus, made a series of observations of Mars and found that the planet was tilted at an angle of almost 24 degrees on its axis of rotation. This finding showed that like Earth, Mars had seasons; however, considering that a Martian year is almost double that of Earth, its seasons are nearly twice as long. Herschel also confirmed the existence of Mars's polar caps, and postulated correctly that they were composed of ice. Finally, Herschel found that the planet had "a considerable but moderate atmosphere."[14]

Canals on Mars

In 1877, Mars came to a perihelic opposition just 35 million miles from Earth. That year Asaph Hall, director of the U.S. Naval Observatory, turned that institution's 26-inch refractor telescope toward the red planet in search of satellites. In August, he discovered two small moons orbiting Mars, which he named Phobos (fear) and Deimos (flight)—these were Mars' attendants in Homer's *Iliad*. Hall continued his observations for several months, using the data he acquired to make an estimate of the mass of Mars. His calculation of 0.1076 times that of Earth proved to be quite accurate (the current accepted value being 0.1074).[15]

---

[12] Ibid.

[13] Sheehan, *The Planet Mars*, pp. 9-15; Wilford, *Mars Beckons*, pp. 3-17.

[14] Sheehan, *The Planet Mars*, pp. 16-22; Wilford, *Mars Beckons*, pp. 3-17.

[15] Sheehan, *The Planet Mars*, pp. 23-30.

Mars Wars

While the discovery of two Martian moons was a significant astronomical finding, it was not the only important study of the planet that year. In Italy, the director of the Milan Observatory, Virginio Schiaparelli, spent the summer observing Mars with a fairly small, 8-inch telescope. During his study, he saw what he believed to be faint linear markings on the planet. His maps of the planet showed dark areas seemingly connected by a large system of long, straight lines. Schiaparelli called these lines *canali*, which in Italian means "channels" or "grooves." However, another meaning of the word is "canal," which seemed to indicate that intelligent beings may have constructed a water transport system on Mars. Schiaparelli himself tried to caution against jumping to this conclusion, but his observations fired the public's imagination. As a result, French astronomer Camille Flammarion was justified in stating "[Schiaparelli's] observations have made Mars the most interesting point for us in the entire heavens."[16]

Nearly two decades later, an American named Percival Lowell began his famed observations of Mars. A Lowell biographer wrote that "of all the men through history who have posed questions and proposed answers about Mars, [he was] the most influential and by all odds the most controversial."[17] An amateur astronomer with a gift for mathematics, Lowell plunged into the field aspiring to complete Schiaparelli's earlier work. Using an inherited fortune, he constructed the Lowell Observatory (which had 18-inch and 12-inch telescopes) in the Arizona mountains near Flagstaff. During the summer and fall of 1894, Lowell studied Mars every night with unbounded enthusiasm. His maps of the planet displayed 184 canals, twice as many as Schiaparelli had portrayed. As a result of his observations, he announced to the world that there were indeed canals on Mars constructed by intelligent beings. In 1895, he published *Mars*, within which he vividly described his theories regarding the Martian canals and their builders.[18]

During the coming years, Lowell continued his observations of Mars. With each subsequent opposition, he became increasingly convinced that intelligent beings lived on Mars and had built the canals. Lowell also postulated that the shrinking of the white polar caps and the expansion of darker regions (which he believed to be vegetation) during the Martian summer indicated seasonal renewal. Despite his grand pronouncements, most astronomers were not convinced that his theories had any merit. Their criticism of Lowell was bolstered by the fact that many other astronomers, including Edward Barnard, had studied Mars with far more powerful telescopes and found no evidence of canals. Barnard wrote "I see details where some

---

[16] Wilford, *Mars Beckons*, pp. 23-24.

[17] William Graves Hoyt, *Lowell and Mars* (Tucson: University of Arizona Press, 1976), p. 12.

[18] Wilford, *Mars Beckons*, pp. 24-30.

of his canals are but they are not straight lines at all." It is now believed that Lowell's canals were simply optical illusions produced because the human eye attempts to arrange scattered spots into a line. Despite the eventual erosion of his theories, however, there is little doubt that Lowell's declarations about extraterrestrial Martian life led to greatly increased public interest in the red planet.[19]

Mars in Popular Culture

In 1898, just three years after Percival Lowell popularized the vision of a Mars threaded by canals and peopled by ancient beings, the first great Martian science fiction book was published. The *War of the Worlds*, written by H.G. Wells, is hailed as the greatest alien invasion story in history. The book began with a Martian assault just outside of London. While the Martians at first seemed helpless in the heavy Earth gravity, they quickly exposed their advanced technology in the form of huge death machines that began destroying the surrounding countryside, forcing the evacuation of London. The saving grace for the badly overmatched humans turned out to be common bacteria that the Martians had no immune system to fight off. In 1938, the book was famously adapted for radio by Orson Welles. The retelling of the story, portrayed as a news program about a Martian landing in rural New Jersey, was so believable that millions of Americans actually thought that Earth was being invaded.[20]

Starting in 1917, author Edgar Rice Burroughs began a highly popular series about Mars exploration with the publication of *A Princess of Mars*. In subsequent years, he wrote ten more books tracking the adventures of Captain John Carter on Mars. The series was first published as a longer sequence of serials printed in *All-Story Magazine,* which represented a common strategy for the publication of science fiction novels during that period. The Carter books were considered to be more fantasy than hard science fiction, which was exhibited by the lack of detail regarding how Carter actually got to the red planet—he was magically taken there in the book.[21]

During the Great Depression and the Second World War, there was a conspicuous absence of popular books regarding Mars exploration. The lull was broken when author Robert Heinlein wrote *Red Planet*. Published in 1949, the book followed teenager Jim Marlowe, his friend Frank, and his Martian "roundhead" pet Willis on their travels across the planet to warn a human colony that was the target

---

[19] Ibid.
[20] H.G. Wells, *War of the Worlds* (1898; reprint, New York: Tor, 1986).
[21] Edgar Rice Burroughs, *A Princess of Mars* (1917; reprint, New York: Ballantine Books, 1990).

Mars Wars

of a conspiracy by the Martians.[22] A year later, Ray Bradbury authored his famous book entitled *The Martian Chronicles*. The book was actually a compilation of relatively unrelated short stories about an ancient, dying Martian race. Along with Heinlein's *Red Planet*, the book borrowed heavily from the observations and theories of both Schiaparelli and Lowell—planetary canals were a central accomplishment of the Martian civilizations in both books. These were early examples of how scientific research pushed science fiction novels.[23] In 1956, Robert Heinlein wrote *Double Star*, the most critically acclaimed Martian novel during this time period. The book, which won the Hugo Award,[24] centered on the emotional predicament of an out of work actor, Lorenzo Smythe, who was asked to stand in for an important politician who had been kidnapped. His trouble began when he was forced to take part in an important ceremony on Mars despite the fact that he hated Martians. The book was an interesting rendering of the civil rights struggle going on in the United States at the time.[25]

During this same period, a large number of popular films featured adventures involving the red planet. In 1938, *Flash Gordon: Mars Attacks the World* premiered as a feature-length film. In the movie, Flash Gordon blasts off for Mars to destroy a mysterious force sucking the nitrogen from Earth's atmosphere and foil a plot by Ming the Merciless to conquer the universe. This was followed in the post-War period with the 1950 film *Rocketship X-M*, the story of five astronauts that set off to explore the moon but due to a malfunction ended up on Mars—where they find evidence of an advanced civilization nearly destroyed by an atomic holocaust. The next year, *Flight to Mars* chronicled the adventures of a team of scientists and a newspaper reporter that fly to Mars and thwart a plan by the Martians (who look identical to humans) to conquer Earth. In 1953, *Invaders from Mars* told the tale of small town where all the adults begin acting strangely shortly after young David MacLean sees strange lights settling behind a hill near his home. That same year, Gene Barry starred in a film adaptation of *War of the Worlds*. Finally, in 1959, *The Angry Red Planet* followed a group of astronauts that land on Mars and battle aliens, a giant amoeba, and the dreaded "Rat-Bat-Spider thing." By the late-1950s, the combination of these best selling books and feature-length films had fixed human exploration of the Mars (and the likely inhabitants of that planet) in the popular culture of the nation.

---

[22] Robert Heinlein, *Red Planet* (New York: Ballantine Books, 1949).

[23] Ray Bradbury, *The Martian Chronicles* (New York: Doubleday, 1950).

[24] One of the two most prestigious awards for accomplishments in science fiction—the other is the Nebula Award.

[25] Robert Heinlein, *Double Star* (New York: Ballantine Books, 1956).

Chapter 1: Introduction

Mariner and Viking

While there was substantial progress made in telescope technology during the 70 years after Lowell's sensational observations, it was still beyond the abilities of astronomers of the time to unequivocally disprove his theories. In fact, during this period there was little sustained interest in planetary astronomy, and as a result, few new discoveries were made. In 1957, the Soviet launch of *Sputnik* opened vast new opportunities for scientific investigations. Once the concept of robotic planetary exploration was conceived during the coming years, it was taken for granted that missions to Mars would be a priority. Several failed attempts by both the Americans and Soviets to send spacecraft to Mars during the early 1960s, however, delayed the first close up examination of the red planet.[26]

On 28 December 1964, NASA launched Mariner 4 on a mission to explore Mars. About halfway to the planet, the spacecraft experienced technical difficulties that greatly concerned ground controllers. The "Great Galactic Ghoul,"[27] however, was unsuccessful in its efforts at crippling the probe. On 14 July 1965, Mariner 4 made a flyby to within 6,118 miles of the planet's surface. It was able to relay 22 images back to Earth with its single camera before passing out of range. The data that was obtained from those images, as well as from the spacecraft's other instruments,[28] were nothing less than stunning. Instead of the living planet that Lowell had envisioned, Mariner 4 discovered a surface that was apparently devoid of life and seemingly unchanged for billions of years. In addition, results of an S-band radio occultation experiment found that the Martian atmospheric density was considerably lower than expected and that its makeup was approximately 95% carbon dioxide. Finally, it was discovered that the planet had no discernible magnetic field. The information returned by *Mariner 4* resulted in a complete revision of human thinking about Mars, ending forever Lowellian theories regarding vegetation and intelligent beings.[29]

During the early months of 1969, the Americans and the Soviets each sent two more spacecraft towards Mars.[30] While the Soviets continued their string of failures, both Mariner 6 and Mariner 7 were successful. These spacecraft, like Mariner 4, were designed as flyby missions, but they were capable of photographing the planet

---

[26] Wilford, *Mars Beckons*, pp. 53-56.

[27] A myth developed by flight engineers after earlier missions to Mars failed, which lives on today.

[28] Including a magnetometer and a trapped-radiation detector.

[29] Sheehan, *The Planet Mars*, chapter 11.

[30] About every 26 months, Mars and Earth reach a position in their respective orbits that offers the best trajectory between the two planets. During this time period, the *Mariner* missions were launched to take advantage of these launch windows.

Mars Wars

at much greater distances. Mariner 6 sent 75 images earthward, while Mariner 7 produced 126 photographs. In total, the two probes, which passed within 2,120 miles of the planet, returned data about approximately 20% of the surface. Once again, the information obtained showed a largely cratered landscape, although it also showed large expanses that were like an exceedingly dry and cold desert.[31]

As chance would have it, the first three Mariner missions explored some of the most geographically lackluster areas of Mars. Launched on 30 May 1971, Mariner 9, the first successful orbiter to reach Mars, finally revealed the topographical diversity of the red planet. When the spacecraft arrived in November, however, the planet was obscured for weeks by a massive dust storm. Two Soviet landers, Mars 2 and Mars 3, were lost in the storm, because they were not capable of waiting in orbit for it to clear. They did, however, become the first machines to reach the Martian surface. A month after Mariner 9 reached orbit, the dust finally cleared, and it was able to begin mapping the planet. The first features that were discovered were a series of gigantic shield volcanoes—the largest being Olympus Mons, the largest known mountain in the solar system. The second major finding was the immense Valles Marineris system, which dwarfed the Grand Canyon and stretched one-quarter of the way around the planet. Finally, the spacecraft detected wide channels (reminiscent of river valleys) and the hummocky terrain that is characteristic of the south polar regions. In October 1972, when the probe ran out of fuel, it had taken 7,239 photographs and revealed a truly unique planet.[32]

After the success of the Mariner program, the next step in the exploration of Mars involved sending robotic vehicles to conduct *in situ* experiments. In the late summer of 1975, Viking 1 and Viking 2 were launched to the red planet to carry out a search for Martian life, among other scientific objectives. Each spacecraft actually had two separate components—an orbiter based on Mariner 9 technologies and a lander equipped with various scientific instruments. On 20 July 1976, about a month after it had entered orbit and seven years after the first human landing on the moon, the 1,300-pound Viking 1 lander settled onto the western slopes of Chryse Planitia—it was the first probe to safely reach the planet's surface. The lander quickly began photographing its surroundings, including a stunning 300-degree panorama that showed sand dunes, a large impact crater, low ridges, scattered boulders, and a pink sky.[33]

The Viking 1 lander was outfitted with a large array of sophisticated equipment, including: antennas for communicating with ground controllers on Earth; cam-

---

[31] Wilford, *Mars Beckons*, pp. 60-61.
[32] Sheehan, *The Planet Mars*, chapter 12.
[33] Wilford, *Mars Beckons*, pp. 86-90.

Chapter 1: Introduction

eras capable of transmitting photographs in black and white, color, and infrared; a mechanical arm capable of scooping soil for examination; and a meteorology boom for assessing atmospheric humidity, temperature, and wind speed. Eight days after landing, the mechanical arm went into action and scooped up its first sample of Martian soil. The soil was released through a funnel that automatically separated it for chemical and biological analysis. While the findings of the lander's various experiments were initially ambiguous, it is the widely held opinion of most of the scientific community that they revealed no signs of Martian life.[34] The Viking 2 lander, which touched down on Utopia Planitia on 3 September 1976, similarly

First Viking 1 panoramic photograph of Martian surface
(Courtesy NASA/JPL-Caltech, Image #PIA00383)

revealed a Mars with no visible signs of living organisms. Despite the conclusions drawn by mission scientists that Mars was lifeless, however, there is still active debate regarding the possibility that the red planet once harbored life. When the durable Viking 1 lander finally ceased operations in November 1982, the first phase of robotic exploration of the planet officially came to an end. Although many of the beliefs that had endured during the first two-thirds of the 20th century had been disproved by robotic probes, there remained considerable interest in future journeys to Mars. The question at that time was whether this second phase of discovery would be centered on robotic or human exploration.

---

[34] Some astrobiologists believe that the Viking Lander's Labeled Release (LR) experiment proved that primitive life does exist on present-day Mars. The LR experiment dropped liquid nutrients onto a sample of Martian soil, then measured the gases that were released by the mixture. If Martian bacteria had consumed the nutrients and had begun to multiply, certain gases would have been released. When the LR experiment was conducted on both Viking Landers, some of the gases emitted seemed to suggest that microbes were ingesting the released nutrients. But, overall, the results were ambiguous. Many in the scientific community believe that the LR results can be explained non-biologically. One such explanation is that the LR experiment showed the surface of Mars to contain oxides. When the nutrients mixed with the oxides, a chemical reaction, not a biological one, occurred. Moreover, these oxides would actually prevent life from forming on the Martian surface. This remains an open debate within the scientific community, although the prevailing belief is that the Viking LR readings did not provide evidence of life on Mars. [Staff Writer, "The Viking Files," *Astrobiology Magazine* (29 May 2003)]

The ensuing chapters will examine the events leading up to the announcement of SEI, including an effort by NASA to garner political support for a crewed mission to Mars during post-Apollo planning. The central focus of this story, however, will be a detailed account of the agenda setting process that placed SEI on the government agenda and the intense political battles that virtually guaranteed that an actual program would not be adopted. Finally, the manuscript will investigate the lessons learned from this failed policy process in an effort to provide a tool to current and future policy makers attempting to garner continued political and public support for human exploration beyond Earth orbit.

2

The Origins of SEI

"Mars responds to a fundamental need in all of us. There is a human imperative to explore. People must explore because they are human beings with a desire to expand the scope of human experience. Exploration adds to our knowledge, satisfies our curiosity, and responds to our sense of adventure. We are going to Mars because we are alive, and because it reflects something very special inside each and every one of us."

NASA Associate Administrator, Arnold Aldrich, 1 May 1990

During the four decades prior to President Bush's announcement of SEI, sending humans to Mars had often captured the imagination of the space community as the ultimate 20th century goal for the space program.[1] Throughout that time period, inspired engineers generated scores of sophisticated mission architectures for accomplishing this objective. During the early Nixon administration, NASA's leaders proposed exploration of the red planet as the post-Apollo goal of the American space program. This effort was thwarted, however, by powerful political and budgetary forces. In the early 1980s, a group of enthusiasts held several conferences aimed at reviving interest in exploration of Mars. This campaign, combined with the recommendations of two important advisory committees, resulted in the

---

[1] Howard E. McCurdy, *Space and the American Imagination* (Washington, DC: Smithsonian Institution Press, 1997).

Mars Wars

Reagan administration officially placing human exploration beyond Earth orbit on the space agenda. This chapter highlights the 40-year "softening up" process that laid the foundation for President Bush's announcement of SEI in the summer of 1989. This historical background will set the context under which human exploration of Mars ultimately reached the government agenda. More important, it will provide insights regarding emergent trends that increased the likelihood that Mars exploration would receive favorable consideration within important parts of the space policy making community, but also should have forewarned key policy makers that there were great challenges to adopting a costly new human spaceflight program from other parts of that community.

Early Mission Planning

In 1952, Dr. Wernher von Braun[2] published the first detailed mission architecture for human exploration of the red planet in his classic book, *The Mars Project*. The manuscript was actually the appendix of an earlier, unpublished work that von Braun had written while interned with his fellow German rocket engineers in El Paso, Texas after the conclusion of World War II (WWII). Von Braun had a sweeping vision for human travel to Mars. His plan called for ten 400-ton spacecraft capable of transporting a crew of 70 to the red planet—almost 1,000 ferry flights would be required to assemble this massive "flotilla" in Earth orbit. The strategy incorporated a minimum-energy trajectory that would carry the ships to Mars in approximately eight months. Upon arriving in Martian orbit, a glider would descend for a sliding landing on one of the planet's polar ice caps. The crew from that ship would then trek 4,000 miles to the equator to build a landing strip for two additional gliders, which would transport the remainder of the exploratory team to the surface. After setting up an inflatable base camp habitat, the crew would commence a 400-day survey of the planet—which, von Braun assumed, would include taking samples of local flora and fauna and exploring the Martian canals—this was more than a decade before Mariner 4 returned the first close-up images of Mars back to Earth. Following a year of exploration, the crew would return to Earth,

---

[2] German engineer who played a prominent role in all aspects of rocketry and space exploration, first in Germany (he led the V-2 rocket program) and, after World War II, in the United States. After working for the U.S. Army, von Braun became Director of NASA's Marshall Space Flight Center and the chief architect of the Saturn V launch vehicle. He began developing ideas for Mars exploration as early as 1947, while working at White Sands.

Chapter 2: The Origins of SEI

These paintings by Chesley Bonestell illustrate von Braun's plan for human exploration of Mars, from construction of the spaceships in Earth orbit, to entering Mars orbit, to exploring the surface itself (Courtesy Chesley Bonestell archives)

completing its three-year journey.[3]

In the early 1950s, *Collier's* magazine approached Wernher von Braun and several other prominent engineers and scientists with an offer to write a series of eight articles about space exploration. The publication of these articles in *Collier's*, with its circulation of almost four million, represented the beginning of a concerted "softening up" process for space exploration in general, and Moon and Mars exploration in particular—with the express purpose of educating the American public.[4] In April 1954, von Braun and journalist Cornelius Ryan penned an article for *Collier's* entitled "Can We Get to Mars?" This piece drew heavily from the mission concept found in *The Mars Project*, but also included an Earth-orbiting space station that would be used during the project's construction phase. In addition, Von Braun included a discussion that analyzed the potential physical and psychological difficulties that the astronaut crew would face on the voyage. He concluded, "…we have, or will acquire, the basic knowledge to solve all the physical problems of a flight to Mars… [but] psychologists undoubtedly will [have] to make careful plans to keep up the morale of the voyagers." In 1956, von Braun collaborated with fellow German engineer Willy Ley to expand on the *Collier's* articles in a book entitled *The*

---

[3] Wernher von Braun, *The Mars Project* (Urbana, IL: University of Illinois Press, 1962); Erik Bergaust, *Wernher von Braun* (Washington, DC: National Space Institute, 1976), pp. 153-159; David S. Portree, *Humans to Mars: Fifty Years of Mission Planning, 1950-2000* (Washington, DC: NASA History Division, 2001), pp. 1-4.

[4] John Kingdon defines "softening up" as a process to pave the way in preparation for opening a policy window. In this process, the policy entrepreneur must ask who must be softened up: the general public, some specialized public, or the policy community itself. Among the means of softening up or educating is conducting and releasing studies or reports relating to the policy, which was the method chosen by von Braun (and others) during the 1950s. [John W. Kingdon, *Agendas, Alternatives, and Public Policies* (New York: HarperCollins College Publishers, 1995), pp. 127-131.]

Exploration of Mars. The manuscript introduced a refined, cheaper mission architecture that reduced the number of ships going to Mars from 10 to 2, and the number of crew from 70 to 12.[5]

In the mid-1950s, the *Collier's* articles served as the basis for three animated films about space exploration produced by Walt Disney. Wernher von Braun served as technical advisor for the shows, while Disney provided artistic direction for the series. The American Broadcasting Company aired the first episode entitled "Man in Space" on 9 April 1955. Disney, introducing the broadcast, stated that the aim of the series was to merge "the tools of our trade with the knowledge of the scientists to give a factual picture of the latest plans for man's newest adventure." The episode introduced fundamental scientific principles and concluded with von Braun's vision for a four-stage orbital rocket ship. The second show, entitled "Man and the Moon," aired on 28 December 1955 and "present[ed] a realistic and believable trip to the moon in a rocket ship—not in some far-off fantastic never-never land, but in the near foreseeable future." The final show in the series, entitled "Mars and Beyond," aired almost two years later on 4 December 1957. During this episode, von Braun and Dr. Ernst Stuhlinger revealed plans for "atomic electric space ships [that] feature[d] a revolutionary new principle that will make possible the long trip to Mars with only a small expenditure of fuel." The Disney technicians provided dramatic animations of a 13-month voyage employing these nuclear rocket engines. *TV Guide* stated that "Mars and Beyond" represented "the thinking of the best scientific minds working on space projects today, making the picture more fact than fantasy."[6]

Howard McCurdy argues in his book *Space and the American Imagination* that von Braun's collaborations with *Collier's* and Disney were part of a larger concerted effort to prepare the public for the inevitable conquest of space. He contends that scientists, writers, and political leaders sought to construct a romantic vision of space exploration laid upon images already rooted in the American culture, such as the myth of the frontier. The resulting vision of space exploration had the power to excite, entertain, or frighten (i.e. Cold War)—and it was incredibly successful. In 1949, only 15% of the population believed that we would go to the moon in the

---

[5] Portree, *Humans to Mars*, pp. 1-4; Bergaust, *Wernher von Braun*, pp. 53-59; Wernher von Braun with Cornelius Ryan, "Can We Get to Mars?" *Colliers*, 30 April 1954, in *Exploring the Unknown: Organizing for Exploration,* ed. Logsdon, et al. (Washington, DC: NASA History Division, 1995, NASA SP-4407, Volume 1), pp. 195-200; Willy Ley and Wernher von Braun, *The Exploration of Mars* (New York: Viking Press, 1956).

[6] Mike Wright, "The Disney-Von Braun Collaboration and Its Influence on Space Exploration" in *Selected Papers from the 1993 Southern Humanities Conference,* ed. Daniel Schenker, Craig Hanks, and Susan Kray (Huntsville, AL: Southern Humanities Press, 1993).

20th century. By the time President Kennedy announced the lunar landing goal, however, the majority of Americans viewed it as inevitable. McCurdy asserts that the primary reason for this shift in national mood was the introduction of space concepts to the mainstream public by von Braun and other visionaries during the 1950s.[7]

Building on the clear rise in interest in space exploration following the launch of *Sputnik* (and solidified with President Kennedy's decision to send humans to the Moon), NASA began a series of studies to investigate alternative mission profiles for sending humans to Mars—continuing the softening up process that Wernher von Braun had initiated in the early 1950s and beginning a long-term alternative generation process within the policy stream. In April 1959, Congress approved funding for NASA Lewis Research Center in Cleveland, Ohio to conduct the first official architecture study for human exploration of Mars. Under the Lewis plan, a crew of seven would be propelled toward the red planet by an advanced, high-thrust nuclear rocket engine. The strategy called for a 420-day round trip with a 40-day surface stay. The ship design provided substantial living space for the crew and a heavily shielded cylindrical vault in the hub to protect the crew from radiation exposure. This basic model, using nuclear propulsion for the Earth orbit to Mars journey, became the standard within NASA for the next decade.[8]

In 1961, the year in which humans first reached Earth orbit, NASA was largely focused on mission plans for sending humans to explore the surface of the moon—not Mars. However, there was at least one important study of Mars exploration that was produced that year. Authored by Ernst Stuhlinger, who directed advanced propulsion work at Marshall Space Flight Center (MSFC), the study expanded on earlier designs for ion-powered spacecraft. This form of propulsion used an electric current to convert a propellant's (e.g. cesium) atoms into positive ions. The engine would then expel these high-speed ions to create a constant low-thrust acceleration. The primary benefit of this vehicle type was that it used relatively little propellant, drastically reducing the amount of launches required to assemble a ship in Earth-orbit. The main drawback, however, was that the low-thrust vehicle would take longer to make the trip to Mars and back. Stuhlinger also introduced a new innovation, twirling the spacecraft to generate artificial gravity for the crew. His overall mission plan called for five 150-meter long twirling ion ships to take 15 astronauts on the voyage to the red planet.[9]

---

[7] McCurdy, *Space and the American Imagination*, pp. 29-82.

[8] Portree, *Humans to Mars*, pp. 5-6.

[9] Portree, *Humans to Mars*, pp. 6-8; David S.F. Portree, "The Road to Mars…Is Paved With Good Inventions," *Air & Space*, February/March 2000, pp. 67-71.

In mid-1962, MSFC commenced the Early Manned Planetary Roundtrip Expeditions (EMPIRE) study. Wernher von Braun, now director of Marshall Space Flight Center (MSFC) and the leading advocate pushing for exploration of Mars, recognized that his field center would need a post-Apollo goal if it were to survive after completion of the Saturn V rocket program. The goal of the EMPIRE study was to provide a long-term human exploration strategy. The study participants were tasked with creating mission plans that utilized moderate modifications of Apollo technology for Martian flyby and orbiter (but not landing) missions. Three EMPIRE contractors submitted reports to MSFC—Lockheed, Ford Aeronutronic, and General Dynamics. The Lockheed and Aeronutronic teams focused primarily on 18 to 22 month flyby missions conducted by spacecraft that utilized a rotating design to create artificial gravity for the crew. The General Dynamics report, on the other hand, focused on orbiter missions conducted by convoys of modular spacecraft. Krafft Ehricke, the principal author of the General Dynamics study, also included options for landing missions. All of the missions proposed under the auspices of the EMPIRE study required launch vehicles capable of lifting 2 ½ to 5 times the weight of the Saturn V being developed for Project Apollo.[10]

In 1963, at the same time MSFC was conducting the EMPIRE study, the Manned Spacecraft Center (MSC) (later renamed the Johnson Space Center—JSC) started to conduct its own advanced planning for the future of the space program. Assistant Director of MSC, Maxime Faget, favored a phased exploration approach, with a space station and lunar base preceding a human mission to Mars. The MSC study produced the first detailed designs for a Mars Excursion Module (MEM), a piloted craft that would be capable of landing on the Martian surface. The mission plan developed by MSC called for a complicated flyby-rendezvous where two separate spacecraft would be sent toward Mars—Direct and Flyby. The Flyby ship would depart Earth on a 200-day trip to Mars. The piloted Direct ship would leave 50 to100 days later on a 120-day trip to Mars. Upon arrival, the Direct ship would release the MEM, which would land on the red planet. After completing its mission, the MEM would rendezvous with the Flyby ship as it swung past Mars and headed home. This high-risk approach saved propellant because it utilized a free return trajectory to return the MEM to Earth.[11]

In July 1965, Mariner 4 conducted the first flyby of Mars and snapped its historic 21 images of the red planet. The photographs revealed a planet with an exceptionally thin carbon dioxide atmosphere and an apparently lifeless, cratered landscape. These findings had a dramatic impact on planning for planetary explora-

---

[10] Portree, *Humans to Mars*, pp. 11-22; Portree, *The Road to Mars*, pp. 67-71.

[11] Ibid.

tion. Researchers had always assumed Mars to be an Earthlike planet that would support a human crew. Instead they found a resoundingly hostile environment. As a result, the next major Mars study, conducted the following year by the Office of Manned Spaceflight at NASA Headquarters, called for humans to orbit the planet but counted on robotic landers to conduct actual surface exploration. By 1967, with the Vietnam War heating up, Congress eliminated all funding for studying human exploration of the red planet.[12]

Post-Apollo Planning

On 8 January 1969, President-elect Richard Nixon received the "Report of the Task Force on Space," a thirteen-member blue-ribbon panel charged with advising the incoming president regarding options for the American space program. Chaired by Nobel Prize winner Charles Townes of the University of California at Berkeley, the task force issued a number of recommendations. The board favored a more balanced program that promoted expanded utilization of robotic probes and satellites for scientific research and exploration, and in a wide variety of applications (e.g., communications, weather, and earth resource surveys). With regard to planetary exploration, the task force did not support the immediate adoption of a human spaceflight program based on a planetary lander or orbiter. Instead, the panel favored continued lunar exploration that built on Apollo technology to allow for greater mobility and extended stays on the surface.[13]

The following month, President Nixon asked Vice President Spiro Agnew to chair a Space Task Group (STG) created to provide a definitive recommendation regarding the course the space program should take during the post-Apollo period. The other members of the STG were Secretary of the Air Force Robert Seamans, NASA Administrator Thomas Paine, and Presidential Science Advisor Lee DuBridge. Joan Hoff argues in *Spaceflight and the Myth of Presidential Leadership* that the creation of the STG was "a mixed blessing for NASA because Paine assumed almost immediately that Agnew's personal and public support of a 'manned flight to Mars by the end of this century' would carry the day inside the White House and BOB [Bureau

---

[12] Ibid.

[13] Charles Townes, et al., "Report of the Task Force on Space," 8 January 1969, in *Exploring the Unknown: Organizing for Exploration*, ed. Logsdon, et al. (Washington, DC: NASA History Division, 1995, NASA SP-4407, Volume 1), pp. 499-512.

Mars Wars

of Budget], when nothing could have been further from the truth."[14,15] At an early STG meeting, Paine pushed forward based on this incorrect assumption by contending that the space agency needed a new program to rally around. Agnew was supportive, stating that NASA needed an "Apollo for the seventies."[16]

As the primary policy entrepreneurs supporting a human mission to Mars, Paine and Agnew selected an approach for post-Apollo planning that did not mesh with either President Nixon's basic ideology or changes in the national mood regarding space exploration. Hoff writes "Nixon was concerned about scientific-technological programs that might stress engineering over science, competition over cooperation, civilian over military, and adventure over applications…[and his] emphasis on frugality in government spending prompted caution on his part in endorsing any effort in space." Public sentiment toward the space program had also begun to shift, with increasing concerns that the government had misplaced priorities. A Gallup Poll conducted in July 1969, at the time of the Apollo 11 mission, indicated that only 39% of Americans were in favor of U.S. government spending to send Americans to Mars, while 53% were opposed.[17] Thus, Paine and Agnew were pushing for a large new Apollo-like commitment despite the fact that there appeared to be little or no support for such an undertaking within the White House or the mass public.[18]

On 16 July, at the launch of Apollo 11, Vice President Agnew told reporters that it was his "individual feeling that we should articulate a simple, ambitious, optimistic goal of a manned flight to Mars by the end of the century." Up until this point, NASA had been focusing primarily on a large space station as the logical post-Apollo program. The space agency had been unsuccessful in gaining political support for such a program, however, so Administrator Paine decided that it was the appropriate time to make a human Mars mission the center of future planning. On 4 August, Wernher von Braun came to Washington to brief the STG on options for human exploration of Mars by 1982. After the briefing, Paine informed the panel

---

[14] In fact, Agnew had little influence within the Nixon White House; his strong support for an ambitious post-Apollo program was potentially a liability for NASA, not an asset.

[15] Joan Hoff, "The Presidency, Congress, and the Deceleration of the US Space Program in the 1970s," in *Spaceflight and the Myth of Presidential Leadership*, ed. Roger D. Launius and Howard McCurdy (Chicago: University of Illinois Press, 1997), pp. 92-132.

[16] Richard Nixon, "Memorandum for the Vice President, the Secretary of Defense, the Acting Administrator, NASA, and the Science Advisor," 13 February 1969, in *Exploring the Unknown: Organizing for Exploration*, ed. Logsdon, et al. (Washington, DC: NASA History Division, 1995, NASA SP-4407, Volume 1), pp. 512-513.

[17] NASA History Division, Compilation of Historical Polling Data, Excel spreadsheet provided to author.

[18] Hoff, "The Presidency, Congress, and the Deceleration of the US Space Program in the 1970s," pp. 92-132.

that the mission could be accomplished if NASA's budget was increased to $9 to $10 billion by the middle of the decade—at a time when the NASA budget was only $4.25 billion.[19] This seemed to be contrary to President Nixon's fiscal philosophy as well as existing budgetary realities.

As it became clear that the STG was seriously considering an early Mars mission, widespread criticism of such an undertaking emerged. What was most troublesome for NASA was that formerly vigorous supporters of the space program were opposed to large new projects. Senator Clinton Anderson, Chairman of the Senate Committee on Aeronautical and Space Sciences, stated "now is not the time to commit ourselves to the goal of a manned mission to Mars." Representative George Miller, Chairman of the House Committee on Science and Astronautics, said "five, perhaps ten, years from now we may decide that it would be in the national interest to begin a carefully planned program extending over several years to send men to Mars." The *Washington Post* and *New York Times* both questioned the validity of the enterprise, the latter stating that an early crewed Mars mission was scientifically and technically premature.[20]

In the face of growing opposition to a Mars project, Robert Seamans grew concerned that his colleagues were considering recommending a program that had no political support. Seamans, who had been a senior NASA official from 1960 to 1968, argued that the space agency should utilize its capabilities to address "problems directly affecting men here on Earth." He contended that new human exploration initiatives should be deferred until their technical feasibility was determined. Budget Director Robert Mayo, who had observer status within the STG, agreed with this position. He believed that from a budgetary viewpoint an Apollo-like program was not practical in the near-term. Due to a lack of consensus regarding the exact direction that the post-Apollo program should travel, it was decided that the panel would present the White House with several future program alternatives. Presidential advisor John Erlichman demanded that the report not include any politically infeasible goals, such as a human mission to Mars by 1982.[21]

On 15 September 1969, the STG report was submitted to President Nixon. The panel recommended that "this Nation accept the basic goal of a balanced manned and unmanned space program conducted for the benefit of all mankind." To accomplish this goal, the report stressed five program objectives, including:

---

[19] John M. Logsdon, "The Policy Process and Large-Scale Space Efforts," *Space Humanization Series* (1979): pp. 65-79.

[20] Ibid.

[21] John M. Logsdon, "The Evolution of US Space Policy and Plans," in *Exploring the Unknown: Organizing for Exploration*, ed. Logsdon, et al. (Washington, DC: NASA History Division, 1995, NASA SP-4407, Volume 1), pp. 383-386.

- Expansion of space applications
- Enhancement of space technology for national security purposes
- Continuation of earth and space science projects
- Development of a new space transportation capability and a space station
- Promotion of international cooperation in space

Finally, the group recommended that the nation "accept the long-range option or goal of manned planetary exploration with a manned Mars mission before the end of this century as the first target." The STG report did not, however, support an immediate commitment to any particular future program or initiative. Instead, the panel provided President Nixon with several alternatives and left it to him to choose the best option.[22]

On 7 March 1970, six months after the STG report was submitted, President Nixon offered his first official comments on the future course of the space program. In his statement, the president declared that the space program would be less of a government priority during his administration. Nixon rejected the need for a bold new exploration initiative, arguing "many critical problems here on this planet make high priority demands on our attention and our resources. By no means should we allow our space program to stagnate. But—with the entire future and the entire universe before us—we should not try to do everything at once. Our approach to space must be bold—but it must also be balanced."[23] This statement formally ended NASA's attempts to get approval for a mission to Mars and led to the eventual endorsement of the Space Shuttle program. Joan Hoff argues that there were four major reasons for the failure of NASA to gain approval for a bold post-Apollo initiative. First, President Nixon never "need[ed] to use the space program to prove himself able to deal with the Soviets, as Kennedy and Johnson apparently thought they did. NASA administrators and White House science advisors in 1969-72 failed to appreciate this important shift...." Second, the Nixon administration inherited economic problems generated by immense spending related to the Vietnam War and Great Society social programs. Third, an anti-technology mood within the American public forced policy makers to question whether large spending for the space program was a proper allocation of scarce government funding. As a result, the president decided that there was no political downside to supporting budget cuts for

---

[22] Space Task Group, "The Post-Apollo Space Program: Directions for the Future," September 1969, in *Exploring the Unknown: Organizing for Exploration*, ed. Logsdon et al (Washington, DC: NASA History Division, 1995, NASA SP-4407, Volume 1), pp. 522-525.

[23] *Public Papers of the Presidents of the United States: Richard Nixon, 1970* (Washington, DC: US Government Printing Office, 1971), pp. 250-253.

NASA. Finally, and perhaps most importantly, Hoff contended that "institutional obstinacy at NASA when asked to comply with changing government budgeting methods and changing public expectations about the meaning of the space program" led to a deceleration within the space program.[24] John Logsdon agreed with this line of reasoning, writing in *Exploring the Unknown: Organizing for Exploration*, "…the results of NASA's attempt to mobilize support behind the manned Mars objective were, from the Agency's perspective, little short of disastrous…. What happened to NASA plans and the STG report is best viewed, not in terms of NASA 'winning' or 'losing,' but in terms of what happens when an agency's plans are significantly at variance with what political leaders judge to be both in the long-term interests of the nation and politically feasible."[25] The fact that NASA pushed the Mars initiative despite substantial opposition resulted in discussion of sending humans to the red planet being a taboo subject within NASA for the next decade.

Case for Mars

Indeed, there was little public dialogue at all regarding a human mission to Mars in the decade after the rejection of such an undertaking by the Nixon administration. By the late 1970s, however, the goal of human exploration of Mars reappeared within the aerospace community—primarily due to the work of a small group of space enthusiasts that became known as the "Mars Underground." The movement began in 1978, during the quiet period between the Skylab and Shuttle programs. That year, Chris McKay, an astrogeophysics graduate student at the University of Colorado at Boulder, offered an informal seminar on "The Habitability of Mars." Among the roughly 25 participants were fellow doctoral candidates Carol Stoker and Penelope Boston, engineer Tom Meyer, and computer scientist Steve Welch. The study, which concentrated primarily on the examination of potential Martian terraforming,[26] continued for several years.[27]

---

[24] Hoff, "The Presidency, Congress, and the Deceleration of the US Space Program in the 1970s," pp. 93-95.

[25] Logsdon, "The Policy Process and Large-Scale Space Efforts," pp. 74-75.

[26] The process of modifying a planet, moon, or other body to a more habitable atmosphere, temperature or ecology.

[27] Alcestis R. Oberg, "The Grass Roots of the Mars Conference," in *The Case for Mars*, ed. Penelope Boston (San Diego, CA: American Astronautical Society, 1984), pp. ix-xii; Robert Zubrin and Richard Wagner, *The Case for Mars: The Plan to Settle the Red Planet and Why We Must* (New York: The Free Press, 1996), pp. 70-74; Amy Reeves, "Space Oddities: Local Members of the Mars Underground Have Come Up for Air," *The Sun*, 25 August 1999.

In the spring of 1980, McKay and Boston met Leonard David of the National Space Institute at an American Astronautical Society meeting in Washington, DC. After a lengthy discussion regarding Mars exploration, David suggested that the Mars Underground organize a conference to analyze options for near-future human exploration of the red planet. The group of twenty-something graduate students enthusiastically latched onto the idea and began planning the event for the following year. McKay, Stoker, Boston, Meyer, Welch, and Roger Wilson, another University of Colorado graduate student, sketched out the key areas to be investigated, including: propulsion, design, psychology, medicine, finance, life support, and materials processing. As the idea progressed, they began putting together lists of speakers for what they dubbed the "Case for Mars" conference.[28]

In late April 1981, the Mars Underground hosted the first "Case for Mars" conference at the University of Colorado. It was a relatively small conference, with approximately 100 attendees, but to the organizers it was viewed as an important start. Given that no official report on human missions to Mars had been released in a decade, the gathering was largely an organized brainstorming session. Over four days, workshops and presentations were given on a wide variety of topics. The most important outcomes of the conference were "first, that the participants made contact and communicated their ideas to the public, and secondly, [the development of] an approach to begin answering the questions of whether or not a manned Mars mission was a viable option for our space program."[29]

During the four days of the Case for Mars conference, the participants examined not only the technologies required to carry out a future human mission, but assessed the social, economic, and political impacts of such an enterprise. The general consensus of the conference participants was that the exploration and settlement of Mars offered a technically feasible, unifying goal for the American space program in the 21st century. The proceedings stated, "this is not only a natural evolutionary step of space development, but it can be a new symbol of the pioneering spirit of America in the eyes of the public." Mars was seen as a logical next step for the space program because the Martian environment provided resources that could be utilized for *in situ* manufacturing of life support materials. It was assumed at the time that the Space Shuttle would provide cheap space transportation services, resulting in a Mars expedition that would cost less than the Apollo program.[30]

---

[28] Ibid.

[29] Ibid.

[30] Penelope J. Boston, et al., "Conference Summary," in *The Case for Mars*, ed. Penelope Boston (San Diego, CA: American Astronautical Society, 1984), pp. xiii-xxi.

The attendees produced a list of four precursor missions that would be required before attempting a human landing on Mars. First, to identify a suitable base site, a robotic Mars Polar Orbiter would be required to locate water resources to support the crew. Second, high resolution maps were needed to provide topographic and geological data since the base must be in a safe but scientifically interesting location. Third, a sample return mission would be essential to carryout engineering proof-of-concept tests. Finally, a mission to either Phobos or Deimos was included as a potential launching point for extensive exploration of the Martian surface.[31] The participants produced interrelated technology options to be used when designing the mission profile, which ranged from using a modified Space Shuttle External Tank as a Mars transit vehicle to mining the Martian atmosphere for fuel to artificial gravity. In conclusion, the conferees produced a list of surface activities that could be carried out by the astronaut and scientist crew during its stay. This included construction of underground habitats in a region with access to a confirmable water supply, establishment of processing facilities to utilize Martian resources (to provide air, water, fuel, industrial compounds, building materials, fertilizers, and soil), growing fresh food to supplement stored supplies, and conducting scientific research.[32]

The Case for Mars conferences, which continued every three years until the mid-1990s when the Mars Society was created (this advocacy organization now holds an annual conference), were essentially a resurrection of the "softening up" process that had been started by the space community during the 1950s. Each conference built on those preceding it, spending time studying the fundamentals of spaceflight (from payloads to orbital trajectories) and establishing a close-knit community of engineers and scientists enthusiastic about sending humans to Mars. In 1984, the second conference was utilized to design a complete space system architecture for a Mars expedition. More importantly, however, the conference attracted attendees with greater political influence within the space policy community, among them former NASA Administrator Thomas Paine. The following year, Paine was appointed to lead a blue-ribbon presidential committee tasked with making recommendations for the space agency's future.

National Commission on Space

In 1981, after the initial flight of the Space Shuttle, NASA began to formulate plans for its next large human spaceflight program. During the next two years, the space agency laid the foundation for a presidential decision in support of a space

---

[31] Ibid.
[32] Ibid.

station. NASA Administrator James Beggs regularly justified the station as the "next logical step" for the civilian space program. When policy makers outside the space agency inquired what an orbiting laboratory was a step toward, NASA officials answered that there were a great many missions that a space station could support. NASA, however, resisted pressure from President Reagan's Science Advisor, George Keyworth, to link the station with an eventual human mission to Mars. Beggs, remembering the negative outcome of the STG's endorsement of an expedition to the red planet, decided that the timing was not right to associate the space station with such an undertaking.[33]

In 1984, Congress adopted legislation requiring President Reagan to appoint a National Commission on Space to develop a long-term agenda for the American space program. In March of the following year, Reagan chose Thomas Paine to lead a commission that included Neil Armstrong, Chuck Yeager, and UN Ambassador Jeanne Kirkpatrick. The selection of Paine, who had spent the past 15 years arguing in favor of an aggressive space program, almost ensured a report that supported an expansive future for NASA. The 15-member commission, which held public hearings to solicit ideas, worked for over a year to prepare its report—which was completed a few days after the 28 January 1986 Space Shuttle *Challenger* accident. This unfortunate coincidence limited any potential short-term impact the report might have had. In May 1986, Bantam Books published it in a glossy volume entitled *Pioneering the Space Frontier*.[34] Subtitled "An exciting vision of our next fifty years in space," the report of the National Commission on Space was dedicated to the seven astronauts that had died in the tragic *Challenger* disaster.[35] That catastrophe had focused much attention on NASA's shortcomings at the same time the commission was offering a bold new vision for the future of the space program. Despite a skeptical reaction to the study from Congress, the media, and the public, the report had a significant impact on human spaceflight strategic planning in the years after 1986.[36]

The members of the National Commission on Space stated that the primary goal of the study was to provide a rationale that would set the American space program on a path to "lead the exploration and development of the space frontier, advancing science, technology, and enterprise, and building institutions and systems that make accessible vast new resources and support human settlements beyond Earth

---

[33] John M. Logsdon, "The Evolution of US Space Policy and Plans," p. 392.

[34] Ibid.

[35] Francis Scobee, Michael Smith, Judith Resnik, Ellison Onizuka, Ronald McNair, Gregory Jarvis, and Christa McAulliffe.

[36] National Commission on Space, *Pioneering the Space Frontier: An Exciting Vision of Our Next Fifty Years in Space* (New York: Bantam Books, 1986).

orbit, from the highlands of the Moon to the plains of Mars." To achieve those objectives, the commission put forward specific recommendations that outlined a logical approach for the future of the space agency. These proposals supported three overarching national goals for the civilian space program: earth and space science; human exploration and settlement of the solar system; and the development of space commerce.[37]

The section of the report dealing with exploration, prospecting, and settling the solar system set out a coherent phased approach for human spaceflight in the 21st century. The first phase entailed sending robotic probes to discover and characterize resources that could be used for later voyages to Mars. During the second phase, more sophisticated missions would be sent Marsward to obtain and return samples to Earth. The third phase would involve robotic and human exploration of the red planet. During this final phase, permanent Martian outposts would be established to support ongoing exploration. The overall tenor of these recommendations suggested that human extraction of chemical and mineral resources on the red planet would be one of the primary long-term goals of the space program.[38]

To support its bold vision for the future of the space program, the commission recommended the establishment of seven demonstration programs to advance key technologies for expansion into the solar system, including: flight research on aerospace plane propulsion and aerodynamics; advanced rocket vehicles; aero-braking for orbital transfer; long-duration closed-ecosystems (including water, air, and food); electric launch and propulsion systems; nuclear-electric space power; and space tethers and artificial gravity. The report further stated that the most important action the government could take to open the space frontier was to drastically reduce transportation costs within the inner solar system. The group advocated completing a new space transportation architecture—including an aerospace plane, cargo vehicle, and space transfer vehicle—that could replace the Shuttle fleet by the turn of the century. A next generation aerospace plane would be capable of providing flexible, routine, and economical passenger service into low Earth orbit (LEO). A large cargo vehicle would be capable of delivering payloads into LEO at a cost of $200 per pound. Finally, a space transfer vehicle would be "developed to initiate a 'Bridge Between Worlds.'"[39]

The National Commission on Space concluded that following its fifty-year strategic plan for the future of the space program would have three tangible benefits, "'pulling-through' advances in science and technology of critical importance to the

---

[37] Ibid.
[38] Ibid.
[39] Ibid.

Nation's future economic strength and national security…providing direct economic returns from new space-based enterprises that capitalize upon broad, low-cost access to space, and…opening new worlds on the space Frontier, with vast resources that can free humanity's aspirations from the limitations of our small planet of birth."[40] The commission calculated that to accomplish the goals its report advocated, the annual NASA budget would have to increase threefold—to approximately $20 billion a year. John Noble Wilford wrote in his book *Mars Beckons* that *Pioneering the Space Frontier* received a frosty reception because "its far-reaching proposals seemed to bear too much of a resemblance to science fiction to be embraced by political leaders. And the more modest recommendations tended to get lost in the 'Bridge Between Worlds' imagery of Buck Rogers spaceships." During the mid-1980s, the American public was not highly receptive to long-range, costly space endeavors. As a result, both the White House and Congress largely disregarded the report of the National Commission on Space.[41]

Despite the negative reaction to the study, momentum began to build for a presidential decision making exploration of Mars the next objective of the human spaceflight program. One reason for endorsing this goal was the increased mission planning that the Soviets were undertaking to set the stage for an expedition to the red planet early in the 21st century. In the coming year, a dozen major publications advocated setting Mars exploration as the primary future goal of NASA—ranging from the *New York Times* to *The New Republic*. Support for Mars exploration was far from unanimous, however, with prominent space policy experts arguing for more limited programs aimed at better space science, earth science, and a permanent return to the Moon. In the face of these conflicting viewpoints, NASA decided to conduct its own study of options for the future of the space program.[42]

The Ride Report

In 1986, NASA Administrator James Fletcher asked former astronaut Dr. Sally Ride to chair a task force assigned to respond to the National Commission on Space report and to develop focused alternatives for the agency's future. In August of the following year, the committee released its report entitled *Leadership and America's Future in Space*. In its preface the study suggested that in the aftermath of the *Challenger* accident there were two conflicting views regarding the proper course for the

---

[40] Ibid.

[41] John Noble Wilford, *Mars Beckons: The Mysteries, the Challenges, the Expectations of Our Next Great Adventure in Space* (New York: Alfred A. Knopf, 1991), pp. 145-150.

[42] Ibid.

space program. On one hand, many believed that NASA should adopt a major, visionary goal. On the other hand, many judged that the agency was already over-committed and should not take on another major program. The Ride Committee sided with the first group, although it concluded that the space program should not pursue a single visionary initiative to the exclusion of all others. It contended that championing a solitary project was not good science or good policy making, but argued that the space program did need a strategy to regain and retain leadership in space endeavors.[43]

The Ride Report identified four candidate initiatives for study, each bold enough to restore the United States to a position of leadership in space. Those proposals included:

- Mission to Planet Earth: a program designed to obtain a comprehensive scientific understanding of the entire Earth system—particularly emphasizing the impact of environmental changes on humanity
- Exploration of the Solar System: a robotic exploration program designed to continue the quest to understand our planetary system (including a comet rendezvous, a mission to Saturn, and three sample return missions to Mars)
- Outpost on the Moon: a program designed to build upon the Apollo legacy with a new phase of lunar exploration and development, concluding with the establishment of a permanent moon base by 2010
- Humans to Mars: a program designed to land a crew of astronauts early in the 21st century and eventually develop a permanent outpost on the red planet

The panel made clear, however, that the report "was not intended to culminate in the selection of one initiative and the elimination of the other three, but rather to provide four concrete examples that would catalyze and focus the discussion of the objectives of the civilian space program and the efforts required to pursue them."

If the Humans to Mars option was pursued, the report recommended a three-prong exploration strategy. During the 1990s, the first prong would involve comprehensive robotic exploration, concluding with a pair of Mars Rover/Sample Return missions. The second prong would entail utilizing an orbiting space station to perform an assertive life sciences program intended to examine the physiological effects of long-duration spaceflight—the ultimate goal being to decide whether

---

[43] Sally Ride, *Leadership and America's Future in Space: A Report to the Administrator*, (Washington, DC: NASA, 1987).

Mars-bound spacecraft would require artificial gravity. During the final prong, the space agency would "design, prepare for, and perform three fast piloted round-trip missions to Mars. These flights would enable the commitment, by 2010, to construct an outpost on Mars." The panel favored one-year human missions to the red planet, with astronauts exploring the planetary surface for 10 to 20 days. The plan called for slow, low-energy cargo vehicles to precede and rendezvous with the piloted spacecraft in Martian orbit. These cargo ships would take everything needed for surface activities, plus the fuel required for the return trip. The Ride Report indicated that the ultimate goal of the initiative was to recapture leadership in space activities.[44]

While human exploration of Mars received equal footing with the other three initiatives proposed by the committee, the report argued that an expedition to the red planet should not be the immediate goal of the space agency. The committee wrote, "…settling Mars should be our eventual goal, but it should not be our next goal. Understanding the requirements and implications of building and sustaining a permanent base on another world is equally important. We should adopt a strategy of natural progression which leads step by step, in an orderly, unhurried way, inexorably toward Mars." This finding seemed to mesh with the general feeling of top NASA officials. In fact, Administrator Fletcher stated at the time his belief that Americans should return to the Moon before heading on to Mars. On the other hand, supporters of human exploration of the red planet argued that developing a lunar base would utilize resources that should be applied toward a journey to Mars. Thus, in the aftermath of the Ride Report it was still unclear what strategy the American space program should adopt as it neared the 21st century—although the report had provided policy makers with four well-conceived future alternatives.[45]

President Reagan and NASA's Office of Exploration

During its deliberations, the Ride task force recommended that an organization be created to perform systematic planning for the nation's civil space program. In July 1987, Administrator James Fletcher established the NASA Office of Exploration to coordinate the agency's efforts to promote missions to the Moon and Mars. Fletcher appointed John Aaron, a longtime NASA official, as the first Assistant Administrator for the bureau. Among Aaron's first assignments was to conduct a study, building on the Paine and Ride reports, looking at options for the long range

---

[44] Ibid.

[45] Ibid.

human exploration of the solar system. This effort, which involved representatives from seven NASA field centers and five headquarters program offices, continued for more than a year.[46]

On 19 December 1988, the Office of Exploration submitted to Fletcher its first annual report, *Beyond Earth's Boundaries: Human Exploration of the Solar System in the 21st Century*—which was the final product of the office's year-long strategic study. The study team examined two different alternatives for future human exploration. First, the space agency could develop a series of expeditions that would travel from Earth to new destinations in the solar system. Second, the space agency could focus on an evolutionary expansion into the solar system that would concentrate more on permanence and the exploitation of resources. The NASA-wide effort utilized a technique called exploration case-studies, whereby a series of technical and policy "what if" questions were asked to judge the viability of several mission options. *Beyond Earth's Boundaries* examined four specific case studies:

- a round-trip human mission from Earth to the Martian moon Phobos, which would serve as a stepping stone to a landing on red planet
- a direct human mission to the surface of Mars
- establishment of a human scientific research station on the Moon
- a lunar outpost to Mars outpost plan, which emphasized the use of the Moon as a springboard for further exploration of the solar system

The study team concluded that an expedition to Phobos could be a valuable interim step to a human landing on the Martian surface, offsetting some of the uncertainties that the latter mission could encounter. They also found that utilizing the Moon as a springboard for expansion into the solar system had a number of advantages, such as learning to construct habitats, extract and process mineral resources, and operate and maintain exploratory machinery. It was also believed that using the Moon as a fuel depot would appreciably reduce the total Earth launch mass, greatly cutting overall programmatic costs. In the end, the report favored establishment of a scientific research station on the Moon as a logical stepping-stone to both a permanent lunar outpost and a full-up Mars expedition. The study team did not support a "crash" human exploration program, regardless of the alternative chosen by policy makers. Instead, it preferred that NASA conduct long-lead technology and

---

[46] Howard E. McCurdy, *The Decision to Send Humans Back to the Moon and on to Mars* (Washington, DC: NASA History Division, March 1992); Lyn Ragsdale, "Politics Not Science: The US Space Program in the Reagan and Bush Years," in *Spaceflight and the Myth of Presidential Leadership*, ed. Roger D. Launius and Howard E. McCurdy (Chicago: University of Illinois Press, 1997), p. 161.

life sciences research during the 1990s—including the completion of Space Station Freedom. It was contended that this would provide government officials with the requisite data to make a decision before the turn of the century regarding the best alternative for expansion into the solar system.[47]

During the period that the Office of Exploration was conducting its study, work was going on within the Reagan administration to generate a new national space policy. In 1982, the White House had produced a national space policy under the auspices of the National Security Council. That document stated the central role of the Space Shuttle in the national security and civil space sectors.[48] In the interim, however, there had been important changes in the American space program—including the *Challenger* accident, a greater emphasis on commercial applications, and the National Commission on Space report. Throughout the latter half of 1987, a Senior Interagency Group (SIG) for Space conducted a comprehensive review that reflected those and other changes in the policy environment. On 11 February 1988, an unclassified summary of the *Presidential Directive on National Space Policy* was publicly released. The stated goals of the space policy were:

- to strengthen national security
- to obtain scientific, technological, and economic benefits
- to encourage continuing private-sector investment in space related activities
- to promote international cooperation; and, as a long-range goal
- to expand human presence and activity beyond Earth orbit into the solar system

This presidential directive was the first time since Kennedy's May 1961 speech that human exploration beyond Earth orbit formally made it onto the government agenda. To implement this new policy, the document directed NASA to begin the systematic development of technologies necessary to enable a range of future human missions.[49]

---

[47] Office of Exploration, *Beyond Earth's Boundaries: Human Exploration of the Solar System in the 21st Century* (Washington, DC: NASA, 1988); John Aaron, "NASA Press Conference Prepared Statement," 19 December 1988.

[48] National Security Council, "National Security Decision Directive Number 42: National Space Policy," 4 July 1982.

[49] National Security Council, "Presidential Directive on National Space Policy," 11 February 1988.

Chapter 2: The Origins of SEI

Despite the appearance that President Reagan had made a momentous commitment to sending humans beyond Earth orbit, many space policy experts question the strength of the pledge. American University's Howard McCurdy argues that the policy directive was merely a "gesture designed to please NASA bureaucrats and space exploration advocates who were clamoring for an expedition to Mars." George Washington University's John Logsdon contends that for all intents and purposes the policy was meaningless because it committed the administration to no specific new exploration program. Finally, the Congressional Research Service's Marcia Smith makes the case that human exploration outside the Earth system was not actually part of the government agenda during the Reagan administration; it was simply part of the "space agenda." Despite the weak commitment to the proposal, however, the presidential directive did generate further momentum for the adoption a Moon-Mars initiative by the next president.[50]

---

[50] Howard McCurdy interview via electronic-mail, 16 April 1999; John Logsdon interview via electronic-mail, 18 April 1999; Marcia Smith interview via electronic-mail, 19 April 1999.

3

Bush, Quayle, and SEI

"There are moments in history when challenges occur of such a compelling nature that to miss them is to miss the whole meaning of an epoch. Space is such a challenge."

James Michener, 1979[1]

By 1989, the American space program had been in a steady decline for nearly two decades. NASA had failed to find its footing in the years following the triumphs of the Apollo moon landings. During the intervening period, the space agency had become increasingly conservative, risk averse, and bureaucratic. After failing to gain support for a robust human exploration program, the agency had retreated and become an ever more cautious organization. During this time, the space program had no great supporters in the White House, nor great advocates within the Congress. This forced the agency to focus its political energies on protecting its turf (e.g., the Space Shuttle and space station programs) and trying to slow the regular reductions in its annual appropriation. The result was a NASA that hardly resembled the organization that had taken on the Soviet Union on one of the most prominent battlegrounds of the Cold War—an agency that had won a great victory for the United States.

Despite this long interlude, there had been stirrings within the space policy community in recent years that seemed to indicate that a return to glory might be achievable. The National Commission on Space had recommended human exploration of Mars as the appropriate long-term objective of the space program. The American

---

[1] James Michener, testimony before the Senate Committee on Commerce, Science and Transportation, *U.S. Civilian Space Policy: Hearings Before the Subcommittee on Science, Technology, and Space*, 96th Cong., 1st session, 1979.

public had rallied around NASA in the wake of the Space Shuttle *Challenger* accident. President Reagan had placed human exploration beyond Earth orbit back on the space agenda for the first time in two decades. Perhaps most importantly, President-elect George Bush was an outspoken supporter of the space program—perhaps more supportive then any incoming president in the history of the space age. On the larger national stage, however, forces that are more significant were developing that didn't bode well for the adoption of an overly aggressive or expensive new undertaking in human spaceflight. In particular, a struggling economy and rising deficits were placing enormous pressure on the federal budget. This political reality would be the most important constraint facing adoption of an expanded exploration program and attempts to revitalize the national space program. In fact, the situation was so grave that it seriously called into question whether the new president should support such an endeavor at all. Despite the potential hazards, though, only a few short months after taking office, President George Bush and his key space policy advisors decided to champion an ill-defined yet exorbitantly expensive exploration plan—the Space Exploration Initiative.

George Herbert Walker Bush was born in Milton, Massachusetts on 12 June 1912, the second child of Dorothy Walker and Prescott Bush, an investment banker and later Republican Senator from Connecticut. He grew up a member of the Eastern elite. Biographer John Robert Greene writes that Bush's parents were "…members of the genteel class—well educated, well pedigreed, well mannered, and well connected. They were also wealthy.… The world in which the Bush children were raised then was one in which comfort was never an issue, but neither were the constant reminders that that comfort could not be taken for granted. Prescott Bush used his wealth as a safety net for his children. They were expected to go out, earn their own wealth, and do the same." As befitting one of this social standing, George received a private school education—first attending the Greenwich County Day School and then moving to prep school at Phillips Academy in Andover, Massachusetts. On his 18th birthday, George graduated from Phillips Academy and enlisted in the U.S. Navy. Within a year, he received a commission as an ensign and became the youngest pilot in the navy. During World War II, Bush flew 58 combat missions against Japanese forces, survived being shot down on a bombing mission, and earned the Distinguished Flying Cross. Upon returning from the war, he entered Yale University, where he earned a B.A. in economics in only two years and graduated Phi Beta Kappa. Following college, Bush spent the subsequent two decades earning his fortune as an executive in the oil industry.[2]

---

[2] John Robert Greene, *The Presidency of George Bush* (Lawrence, KS: University of Kansas Press, 2000), pp. 11-26.

In 1964, George Bush made his first run for elective office when he challenged incumbent Democrat Ralph Yarborough for the U.S. Senate. Despite a good showing for a Republican in Texas, a Democratic Party united under the leadership of fellow Texan Lyndon Johnson stymied Bush—he only received 43% of the vote in the November election. Two years later, however, in 1966, Bush made a successful run for a congressional seat in Houston, becoming the first Republican to represent that city. One of the few freshman congressional representatives ever selected to serve on the powerful Ways and Means Committee; he was reelected two years later without opposition. In 1970, at the behest of President Nixon, Bush made another run for the U.S. Senate. This time running against conservative Democrat Lloyd Bentsen, who had surprisingly beaten Yarborough in the primary, he lost once again, garnering only 46% of the vote.[3]

On 11 December 1970, President Nixon, who greatly appreciated Bush's willingness to sacrifice a safe House seat to run for the Senate, appointed him to the post of U.S. Ambassador to the United Nations. He served in this capacity for two years, but in early 1973 Nixon asked him to take the reigns of the Republican National Committee (RNC) in the aftermath of the Watergate break-in. In that position, Bush was an early defender of Nixon. When tapes were released that proved the president was guilty of obstruction of justice, however, he changed his stance and strongly lobbied the president to resign. In December 1975, after serving as the American envoy to China for a year, President Ford appointed Bush as the Director of Central Intelligence (DCI). After Jimmy Carter's election, despite a solid record of achievement at the CIA, he became the first DCI to be dismissed by an incoming president-elect. He spent the next four years preparing to contend for the Republican presidential nomination. In 1980, despite winning the Iowa caucuses, George Bush never recovered from his loss to former California governor Ronald Reagan in the New Hampshire primary. After an effort to create a Reagan-Ford "dream ticket" collapsed, the conservative Reagan asked Bush to accept the vice presidential nomination because he was "the most attractive surviving moderate." Biographer John Robert Greene explains, "Bush's major task as Vice President was to be the administration's front man on the road. Between 1981 and 1989, Bush put in 1.3 million miles of travel, visiting the 50 states and 65 different countries."[4]

---

[3] Ibid.

[4] Ibid.

Vice President Bush meets Shuttle *Challenger* families (NASA Image 86-HC-181)

Bush-Quayle 1988

Before being elected as vice president, despite having served as a naval aviator during WWII, George Bush did not exhibit a particular interest in the American space program. As vice president, however, Bush became increasingly involved in space policy making—particularly in the aftermath of the Space Shuttle *Challenger* accident. Bush met with the families of the lost astronauts just hours after the tragedy and was deeply moved by both their sorrow and continued dedication to NASA. Dr. June Scobee, the wife of the mission commander, told him not to let the disaster hurt the agency and to fight to keep the space program alive. On 21 March 1987, at the dedication of the *Challenger* memorial at Arlington National Cemetery, Bush stated, "the greatest tribute we can pay to the *Challenger's* brave crew and their families is to remain true to their purpose, and to rededicate ourselves to America's leadership in space."[5]

In the summer of 1987, NASA Administrator James Fletcher requested a meeting to brief Vice President Bush on issues relating to the American space program.[6]

---

[5] George Bush, "Remarks for the Space Shuttle Challenger Dedication," 21 March 1987, Bush Presidential Records, George Bush Presidential Library.

[6] Although Vice President Bush had no formal role in space policy making, he had been involved in this issue area in the months after the *Challenger* accident. One can also conjecture that the briefing

Chapter 3: Bush, Quayle, and SEI

Specifically, the administrator wanted to talk about the U.S. and U.S.S.R. civil space programs, the status of the Space Shuttle's return to service, and the keys to the agency's future. On 10 August, Fletcher went to the White House to present the agency's vision for the 1990s and beyond. He told Bush that the Soviets were beating the Americans in space because of the former nation's unrelenting expansion of its civil and military capabilities during the prior three decades. The primary examples of that preeminence were the permanently occupied Mir space station and the massive Energia launch vehicle. Fletcher argued that Space Station Freedom and a Shuttle-derived advanced launch vehicle would not be capable of achieving parity with the Soviets until the end of the century. He suggested that this disparity in technical capabilities gave the U.S.S.R. an advantage in robotic and human exploration of Mars, which was the long-term goal of Soviet space efforts.[7]

Administrator Fletcher indicated that four important steps needed to be taken to reverse the current course of American-Soviet competition. First, NASA needed funding to develop a flexible and robust space transportation system to assure access to Earth orbit. Second, several critical enabling technologies needed to be tested, including: artificial gravity, closed loop life support, aero-braking, orbital transfer and maneuver, cryogenic storage and handling, and large scale space operations. Third, the U.S. needed the national will to excel. Finally, policy makers needed to choose a long-term, post-Space Station goal that would provide a strategic focus for the space program. Fletcher argued that human exploration of the solar system would provide this direction. He outlined a plan that would commence with a return to the Moon, establishment of a scientific outpost, and exploitation of lunar resources. Then, in the early 21st century, the space agency would be ready to start sending human missions to explore Mars. The meeting with the vice president was purely informational in nature and was not aimed at gaining an immediate response. The main objective was to lay the groundwork for further lobbying efforts should Bush be elected to replace President Reagan.[8]

On 13 October 1987, speaking at Houston's Hyatt Regency Hotel, Vice President George Bush announced he would make a second run for the presidency. Just weeks after declaring his candidacy, Bush delivered a major policy address on the future of the space program at the Marshall Space Flight Center in Huntsville, Alabama. While space policy was not a hot issue during the election cycle, Bush gave

---

was intended to gain support for NASA programs from the likely Republican nominee in the following year's presidential election.

[7] James Fletcher, "Briefing for the Vice President of the United States," 10 August 1987, Bush Presidential Records, George Bush Presidential Library.

[8] Ibid.

it more attention than any presidential candidate since John Kennedy. He told the crowd gathered at the NASA field center, "In very basic ways, our exploration of space defines us as a people—our willingness to take great risks for great rewards, to challenge the unknown, to reach beyond ourselves, to strive for knowledge and innovation and growth. Our commitment to leadership in space is symbolic of the role we seek in the world." Exhibiting his moderate Republican roots, Bush spoke eloquently of the environmental problems facing humanity and the role that NASA's Mission to Planet Earth (a planned network of earth science satellites) could play in altering "our disastrous course." He also endorsed aggressive launch vehicle developments aimed at replacing the Space Shuttle, assuring access to low-Earth orbit and drastically reducing launch costs. Finally, drawing on his briefing from Administrator Fletcher two months earlier, Bush argued that the nation "…should make a long-term commitment to manned and unmanned exploration of the solar system. There is much to be done—further exploration of the Moon, a mission to Mars, probes of the outer planets. These are worthwhile objectives, and they should not be neglected. They should be pursued in a spirit of both bipartisanship and international teamwork." He stressed the need for international cooperation, stating that the solar system should be explored not as Americans, Soviets, French, or Japanese, but as humans. Despite delivering a relatively optimistic address about the future of the space program, Bush also openly spoke about the budgetary pressures facing the federal government and its impact on the space agency. "We cannot write a blank check to NASA … while our dreams are unlimited, our resources are not, and we must choose realistic missions that recognize these constraints."[9]

During the course of the next year, the country watched one of the nastiest presidential contests in history unfold. After a slow start with a third place finish in the Iowa caucuses, Bush stormed back with a solid win in New Hampshire followed by a sweep of the Super Tuesday primaries on 9 March 1988. In August, the Republican Party unanimously nominated Bush. On the morning of 3 October 1988, while engaged in a chaotic general election campaign against Governor Michael Dukakis of Massachusetts, the nominee was on hand at Edwards Air Force Base for the landing of the Space Shuttle *Discovery*, the first shuttle mission since the *Challenger* accident. That afternoon, at a campaign stop in Redding, California, Bush strayed from his regular stump speech to talk about the space program. He stated that if he were elected president, he would have certain strong notions about the proper priorities for the space program. Once again drawing upon the NASA briefing he had received the previous year, Bush contended that the logical progression

---

[9] George Bush, "Excerpts of Remarks at George C. Marshall Space Flight Center," 29 October 1987, Bush Presidential Records, George Bush Presidential Library.

of the human spaceflight program would involve completion of the Space Station, followed by missions to the Moon and Mars. He stopped short of endorsing the adoption of an exploration plan focusing on the red planet, however, stating that he was "not completely sure Mars is the next place we ought to go, and I want to receive the best thinking on that. But if we decide to go, we'll have to be ready, and a space station is part of that process."

Vice President Bush meets *Discovery* STS-26 crew (NASA Image 88-HC-410)

Still, Bush made it clear that while robotic probes had great benefits he believed the future of the American space program should place an emphasis on human spaceflight. The primary rationale he gave for human involvement was that missions to explore the solar system were "journeys of discovery and daring, and they will lose their impact and their meaning if they are performed only by machines."[10] The following month, George Bush defeated Michael Dukakis—with a supportive president-elect ready to take office, this opened a policy window for adoption of a long-term human spaceflight initiative. The main question facing the space policy community, however, was whether a plan could be formulated that would revitalize the space program without requiring a large infusion of budgetary resources.

Reagan-Bush Transition

After his election, President-elect Bush set upon the task of selecting a White House staff and appointing a cabinet. During the campaign, President Reagan took extraordinary measures to ease the new president's transition into office. This was the first time in six decades that a president of one party was succeeded through election by a president of the same party. James Pfiffner argues in *Presidential Transitions: The Reagan to Bush Experience*, "The fact that the transition to a Bush administration was a 'friendly takeover' [meant] there was no rush, as there would be with any party-turnover transition, to ensure that the opposition political party was out of office as soon as possible."[11] In retrospect, this proved to be a mixed blessing for the White House when it came to appointing a new NASA administrator. Changing the space agency's political leadership was not a top priority for the Bush transition team. Administrator Fletcher had returned to the space program in the aftermath of the *Challenger* accident to get NASA back on its feet. While it was assumed that he would choose to return to the private sector, there was no perceived hurry in finding his replacement. In coming years, this relatively laid-back approach to naming a new administrator would come back to haunt the White House.

In December 1988, President-elect Bush received a report from the National Academy of Sciences and National Academy of Engineering (NAS/NAE) that provided recommendations for determining the proper role for the civil space program with respect to other claims on federal resources. The NAS/NAE team, chaired

---

[10] George Bush, "Excerpts of Remarks at Redding, California," 3 October 1988, Bush Presidential Records, George Bush Presidential Library.

[11] James P. Pfiffner, "The Bush Transition: Symbols and Substance," in *Presidential Transitions: The Reagan to Bush Experience*, ed. Kenneth W. Thompson (Lanham, MD: University Press of America, 1993), pp. 62-72.

by H. Guyford Stever, decided that the most important policy question that Bush needed to address after taking office regarded the future of the space station program. The study found that "a permanently manned space station is needed to maintain a viable manned spaceflight capability for the United States. However, its primary justification is to establish the feasibility of human exploration beyond Earth's orbit." Thus, the report argued that final decisions regarding the pace of station development and its final configuration should be made in the context of long-term goals for the space program. The report contended that attaining space leadership required a civil space program with two structural components. First, NASA must have a base competency in all forms of space activities to ensure the feasibility of a wide-range of large initiative alternatives. These fundamental capabilities included: assured access to space; a respectable space science and Earth remote sensing program; and sufficient funding for advanced technology development. The report estimated that this foundational element would cost $10 billion per year. Second, NASA should adopt a long-term strategy for major human spaceflight initiatives serving scientific, political, cultural, and foreign policy objectives. The report argued that these projects should be funded separately from the base program, primarily to avoid the erosion of crucial capabilities. Potential initiatives that the study team identified were: the space station; a permanent return to the moon; and a human mission to Mars. The report approximated the annual expenditure for these endeavors at between $3 and $4 billion. The NAS/NAE team also suggested that President-elect Bush should take advantage of opportunities offered by cooperative endeavors with other space faring nations, which would provide costs savings and political, scientific, and technical benefits.[12] Although this approach seemed to match the budgetary realities facing the space program, it was not seriously considered by the incoming administration or NASA.

During the transition period, a four-person team was established within NASA and tasked with studying open issues facing the civil space program. Led by Brad Mitchell, the team spent the two months before the Bush inauguration preparing a report for whoever would eventually replace Fletcher.[13] In mid-January, the NASA Transition Office Contact Team submitted its briefing report for the NASA Administrator-Designate, who was expected to be appointed sometime in the late

---

[12] Robert M. White and Frank Press to George Bush, December 1988, Bush Presidential Records, George Bush Presidential Library; National Academy of Sciences and National Academy of Engineering, *Toward a New Era in Space: Realigning Policies to New Realities* (Washington, DC: National Academy Press, 1988).

[13] Ibid.; NASA Transition Office Contact Team, "Briefing Report to the NASA Administrator-Designate," 20 January 1989, Bush Presidential Records, George Bush Presidential Library.

spring. The transition team concluded that the space agency's strategic planning must be carried out, taking into account the Administration's concerns about the budget deficit. Therefore, it found that an efficient priority-setting mechanism was needed to achieve the most value for NASA's budgeted dollars.[14] The briefing report further outlined the major space programs and themes that President Bush spoke about during the campaign. The transition team argued that one of the most important tasks for the new administrator would be to assist the President in prioritizing these programs. The study separated the various space issues into seven categories: policy formulation; exploration and scientific research; aeronautics and technology development; commercial initiatives; international cooperation; national security; and science and mathematics education. With regard to policy formulation, one of the first tasks for the new administration would be the creation of a National Space Council chaired by Vice President Dan Quayle—Congress had mandated this the previous year.[15] The primary duty of this organization would be to "set ambitious goals for a space comeback and re-establish U.S. preeminence in space." With relation to human exploration, the study made clear that the primary near-term objective was to develop an operational space station by 1996. While the report mentioned the Reagan administration's space policy that called for human exploration beyond Earth orbit, it did not specifically suggest whether the Bush administration should make that same commitment. Instead, the transition team discussed programs that would lay the foundation for such undertakings, including: development of an aerospace plane; advancement of pathfinder technologies; and construction of heavy-lift launch vehicles.[16] Beyond setting out the Bush administration's basic ideas about the space program, the hastily prepared Mitchell Report added little to the ongoing debate regarding human exploration beyond Earth orbit. The contact team itself implied that the next NASA Administrator and President Bush should make strategic policy decisions based on more authoritative studies like the Ride Report and *Beyond Earth's Boundaries*. Therefore, as the new Bush administration came to power, it was still unclear what direction it would take with regard to future human exploration of the Moon and Mars.

---

[14] NASA Transition Office Contact Team, "Briefing Report to the NASA Administrator-Designate," 20 January 1989, Bush Presidential Records, George Bush Presidential Library.

[15] In 1988, Congress included language in NASA budget authorization that required the President to establish a National Space Council. The President was required to submit by 1 March 1989 a report that outlined the composition and functions of the Council, which was to employ not more then seven persons (including an executive secretary appointed by the President).

[16] NASA Transition Office Contact Team, "Briefing Report to the NASA Administrator-Designate," 20 January 1989, Bush Presidential Records, George Bush Presidential Library.

Chapter 3: Bush, Quayle, and SEI

The Problem Stream: Providing Direction to a Directionless Agency

On 20 January 1989, George Bush was sworn in as the nation's 41st president. Although space policy was not a pressing issue for the Bush administration during its first few months in office, work did begin relatively early to put a space policy team in place. This specifically took the form of establishing the National Space Council mandated by Congress. On 9 February, President Bush appeared before a joint session of Congress and delivered his first State of the Union address. In the oration, he proposed a $1.16 trillion "common sense" budget that would give attention to urgent priorities, provide investment in the future, attack the budget deficit, and require no new taxes. The space program was prominently placed in the speech. Unlike previous presidents, Bush made the space program a significant part of the government agenda from the beginning of his presidency. Bush stated efforts must be made to extend American leadership in technology, increase long-term investment, improve the educational system, and boost productivity. To facilitate meeting these goals, the new president stated, "I request funding for NASA and a strong space program, an increase of almost $2.4 billion over the current fiscal year. We must have a manned Space Station; a vigorous safe Space Shuttle program; and more commercial development in space. The space program should always go 'full throttle up.' And that's not just our ambition; it's our destiny…." Despite his strong request for an expanded space program, Bush did not specifically declare himself in favor of human exploration beyond Earth orbit.[17]

A week after his state of the union address, Bush received a congratulatory letter from Republican Senator Jake Garn of Utah. In April 1985, Garn flew aboard the Space Shuttle *Discovery* as the first public official to travel in space. A former Navy pilot and Brigadier General in the Utah Air National Guard, he served as a payload specialist on STS 51-D. Garn, a member of the Senate Appropriations Committee, applauded the president's emphasis on federal investments in science and technology as crucial to the nation's economic vitality and future. Garn wrote, "your words on science, space, and technology promise the aggressive and ambitious stance that your administration will take in pursuing these needed investments for American's future." Despite his praise of Bush's leadership, he enunciated the misgivings of the space community resulting from the failure of the administration to move quickly in appointing a leadership team for the space program. He lauded the proposed creation of the National Space Council under the leadership of Vice President Quayle, but suggested that the organization's "membership and staff must be selected and

---

[17] *Public Papers of the Presidents of the United States*, 9 February 1989, *Address on Administration Goals Before a Joint Session of Congress*, http://bushlibrary.tamu.edu/papers/ (accessed 18 May 2002.)

mobilized quickly to make a difference in the current budget cycle." He concluded by saying that strong and decisive leadership was needed to counteract the power of entrenched political interests that resist new budgetary initiatives.[18]

A few weeks later, President Bush responded to Senator Garn's call to action regarding the national space policy agenda. On 1 March, press secretary Marlin Fitzwater announced in a press release the appointment of "Dr. Mark Albrecht as the Director of the staff of the National Space Council which is to be created by Executive Order." The administration's first choice for the position, a State Department official with military space experience named Henry Cooper, had asked that his name be withdrawn after concerns were raised about his ties to former Senator John Tower.[19] Although Albrecht brought a wealth of Washington experience to his post, he was not well known within the civilian space community. "Albrecht was a typical Hill rat, a squat, bearded infighter with a Ph.D. from the RAND Corporation who…knew next to nothing about NASA."[20] His background was entirely within the national security policy arena. For the previous six years, he had served as National Security Advisor to Senator Pete Wilson of California. Prior to that, he had worked on the Strategic Defense Initiative (SDI) as a Senior Research Analyst at the Central Intelligence Agency. In 1988, he was the chief drafter of the defense planks for the Republican Party platform—where he caught the attention of Vice President Bush. Dr. Albrecht's background in the national security space program, with its preference for clean sheet planning processes and requirements-based program development, would be one of several factors that would eventually contribute to a dysfunctional relationship between the Space Council and NASA.

The appointment of a national security and weapons specialist heightened already existing concerns among the space community that White House policy would tilt toward military interests. John Pike from the Federation of American Scientists said, "I can't see that Albrecht brings much to the table besides 'Star Wars.' I'm looking for a silver lining here and haven't seen it yet."[21] Reaction on Capitol Hill was more

---

[18] Senator Jake Garn to President George Bush, 16 February 1989, Bush Presidential Records, George Bush Presidential Library.

[19] Former Texas Sen. John Tower was tapped by President Bush to become defense secretary, but the nomination quickly ran into trouble as opponents questioned Tower's business dealings with defense contractors. The confirmation hearings also brought Tower's personal life squarely into the public eye, with some critics alleging he drank excessively. At one point, Tower pledged to quit drinking entirely if confirmed, but his appointment was rejected 53-47 by the Senate in March 1993.

[20] Bryan Burrough, *Dragonfly: NASA and the Crisis Aboard Mir* (New York, NY: HarperCollins Publishers, 1998), p. 239.

[21] Sharen Shaw Johnson, "Capital Line," *USA Today*, sec. 6A, 3 March 1989.

muted. Congressional staffer Stephen Kohashi remembers that those "…responsible for civilian space activities [weren't] familiar with [Albrecht]. I [recall] learning that his space background was primarily in the military or intelligence world, and being somewhat concerned. [But] I don't recall considering this appointment [to be] particularly critical relative to the effectiveness of the council."[22] Despite any potential misgivings from outside the administration, Albrecht clearly had the support of Vice President Quayle—whom he had periodically done work for as a Senate staffer. In his vice-presidential memoir *Standing Firm*, Quayle wrote that NASA at that time "was, to a great extent, still living off the glory it had earned in the 1960s, and I thought Mark Albrecht was just the sort of guy who could shake it up." It also seemed that Democratic staffers would be willing to work with him. One top aide was quoted at the time as saying, "It's not what Albrecht did before. It's what he does in the future. The important thing is to make this work … to formulate a policy consensus and get some pragmatic policy decisions out of the White House."[23] The selection of Albrecht, who would become one of the key policy entrepreneurs supporting human spaceflight beyond Earth orbit, was a major catalyst to assembling a dedicated staff committed to providing the space program with a new vision for the future. This path, however, would not be without obstacles. As one commentator pointed out, the biggest "challenge facing Albrecht…will be to negotiate peace and find common ground among the competing interests on the [Space Council].… Albrecht is very bright, very competent, but nothing can prepare you for that kind of work. It's like war."[24] It would remain to be seen whether this civil space policy neophyte would be able to control the various elements of the space community, especially NASA.[25]

On 9 March, Administrator Fletcher sent a letter to President Bush that would prove to be an important stepping-stone on the road to the announcement of SEI. The letter addressed the forthcoming 20th anniversary of the first human landing on the moon. Fletcher wrote the president that the occasion provided a unique

---

[22] Stephan Kohashi interview via electronic mail, Washington, DC, 16 November 2004.

[23] Kathy Sawyer, "Concern Rises Over Space Council's Direction," *The Washington Post* (9 March 1989): A23.

[24] Eliot Marshall, "An Arbitrator for Space Policy," *Science* (10 March 1989): p. 1283.

[25] Press Release, White House Office of the Press Secretary, 1 March 1989, Bush Presidential Records, George Bush Presidential Library; Dwayne A. Day, "Doomed to Fail: The Birth and Death of the Space Exploration Initiative," *Spaceflight* (March 1995): pp. 79-83; "National Space Council Director Named, Report Sent to Congress," *Aerospace Daily* (3 March 1989); Sawyer, "Concern Rises Over Space Council's Direction," A23; Dan Quayle, *Standing Firm: A Vice-Presidential Memoir* (New York, NY: HarperCollins, 1994), p. 179.

opportunity to define the administration's commitment to the exploration of space. The administrator suggested that Bush's participation in an event planned at the Smithsonian's National Air and Space Museum (to be attended by Apollo 11 astronauts Neil Armstrong, Buzz Aldrin, and Michael Collins) would enhance the significance of the anniversary. Fletcher suggested that, "Taken by itself, an anniversary of this sort tends to focus on past glories and a nostalgia for days long gone. Coupled with a message of leadership and strong direction for the future, it becomes an integral part of the American space experience; it can reenergize the country by setting new challenges and new horizons in the historic context of earlier goals successfully met...." Having received this letter, National Security Advisor Brent Scowcroft wrote that the President "has made it clear that space policy and exploration would be a priority of his Administration. This would be a tangible demonstration of his commitment." This letter from Fletcher got the White House thinking about the possibility of utilizing the anniversary, only four months away, as a platform for a major space policy speech.[26]

It was clear during the early months of the Bush administration that the president had not settled on Mars exploration as the penultimate goal of the American space program. On 16 March, while speaking at a Forum Club luncheon in Houston, he was asked to comment about his support for the Space Station and a human mission to Mars by the end to the century. Bush replied, "On the Space Station, I am strongly for it. We have taken the steps, budgetwise, to go forward on that. I have not reached a conclusion on whether the next major mission should be a manned mission to Mars... [W]e're asking the [newly reconstituted] Space Council... to come forward with its recommendations. So, no decision is made [regarding] what happens beyond the Space Station itself, and I will make that decision when I get their recommendation." Once again, the president cautioned that although his Administration had requested a budget increase for NASA, constrained resources meant that he was not ready to support the immediate adoption of a human mission to Mars. This statement indicated that just four months before announcing a major new space initiative, neither President Bush nor his senior space policy advisors had committed to a costly new human spaceflight program.[27/28]

---

[26] Letter, James C. Fletcher to President George Bush, 9 March 1989, Bush Presidential Records, George Bush Presidential Library; Memorandum, Joseph Hagin to Brent Scowcroft, 3 April 1989, Bush Presidential Records, George Bush Presidential Library.

[27] *Public Papers of the Presidents of the United States*, 16 March 1989, *Remarks at a Luncheon Hosted by the Forum Club in Houston, Texas,* http://bushlibrary.tamu.edu/papers/ (accessed 19 August 2003).

[28] Eight days later this theme was confirmed when the White House hosted the crew of STS-28,

Chapter 3: Bush, Quayle, and SEI

As the Bush administration was beginning to assemble its space policy team, the Center for Strategic and International Studies (CSIS), a Washington-based think tank, released a report analyzing national space policy. Entitled *A More Effective Civil Space Program*, the study was significant because it had been co-chaired by Brent Scowcroft before his appointment as the new president's national security advisor. The report suggested that while NASA's charter was to help maintain American leadership in science and technology, it was far from clear whether the agency was meeting that objective in the post-Apollo era. CSIS indicated that a combination of factors, including a declining budget and a short-sited planning process, had held the space program back during the prior two decades—while the Soviets, Europeans, Japanese, and Chinese were all making significant strides in their space programs. The task force contended that to reverse this disturbing trend the civil space program should figure more prominently on the national agenda. To accomplish this goal, CSIS recommended that the Bush administration and Congress "set solar system exploration by means of automated and piloted spacecraft as a long-term national objective. This should include an important 'Humans to Mars' component, with deliberate and orderly preparation." The report advocated a gradually planned buildup of key technologies and skills, rather than an accelerated program that overextended current capabilities. Finally, CSIS proposed establishing a two-year study effort aimed at developing programmatic alternatives for implementing this long-term strategy. As Administrator Fletcher had done just weeks earlier, the report concluded that 20 July 1989 would be an auspicious date for announcing a new initiative because of its symbolic importance.[29]

By early spring, with momentum growing for a presidential announcement on the 20th anniversary of the Apollo 11 lunar landing, Mark Albrecht broached the idea of a new human spaceflight initiative with Vice President Dan Quayle. He argued that the adoption of a long-term goal would provide focus to a directionless space program and a means to prioritize and streamline existing programs, espe-

---

which had successfully landed the shuttle *Discovery* the previous week at Edwards Air Force Base in California—the primary task of the mission had been the deployment of Tracking and Data Relay Satellite-4 (TDRS-4). In congratulating the crew, President Bush stated that "the story of *Discovery* is as ... timeless as our history ... it says that to Americans—nothing lies beyond our reach." In his brief remarks, the President reaffirmed his commitment to the shuttle program, space science, and construction of the Space Station Freedom. He did not, however, mention human exploration beyond Earth orbit as one of the goals of his administration. [Press Release, White House Office of the Press Secretary, 24 March 1989, Bush Presidential Records, George Bush Presidential Library.]

[29] John H. McElroy and Brent Scowcroft, *A More Effective Civil Space Program* (Washington, DC: The Center for Strategic and International Studies, 1989).

cially the Space Station. Quayle liked the idea. Although he had not been heavily engaged in space policy during his political career, the new vice president would become the single most important space policy entrepreneur within the Bush White House. Born in 1947, James Danforth Quayle was the grandson of Eugene Pulliam, founder of an empire of conservative newspapers. Although he spent most of his childhood in Arizona, Quayle's political roots were sowed in Indiana. After graduating from DePauw University and Indiana State University Law School in Indianapolis, he was elected to the U.S. House of Representatives at the age of 29. Four years later, he became the youngest person ever elected to the U.S. Senate from Indiana. During his Senate tenure, Quayle focused his legislative work in the areas of national defense, arms control, and labor policy. In August 1988, George Bush tapped Dan Quayle to be his running mate. Although this selection was widely criticized because it was felt that Quayle did not have enough experience to be president should something happen to Bush, these opinions had little impact on the ultimate outcome of the election. As vice president, Quayle became the first Chairperson of the National Space Council, which had been statutorily re-established the previous year by Congress. Quayle embraced this assignment, telling the media, "for the first time in a long time there will be a space advocate in the White House—and that will be me." Within a short period, Quayle came to believe that the Space Council's primary objective was to fix a dysfunctional space agency. The vice president agreed with Mark Albrecht that the main rationale for adopting a new human spaceflight initiative was to restore NASA to prominence. His main concern, however, was achieving this renewal given the existing budgetary and political environment.[30]

Given his unease regarding the budgetary constraints, Quayle immediately scheduled a meeting with Office of Management and Budget (OMB) Director Richard Darman to discuss announcing a new initiative. In his memoirs, Quayle stated, "despite his image as a budget cruncher and arm-twister … Dick Darman has his visionary side, and he was a cheerleader for a manned mission to Mars." This became apparent at a meeting in early April attended by Quayle, Albrecht, Darman, and OMB Executive Associate Director Bob Grady. Darman was supportive, but cautioned that the current budget situation mandated that the administration could not serve up an Apollo-like crash program. As a result, three basic tenets emerged from this meeting. First, recognizing that there were interest groups in favor of both Moon and Mars exploration, President Bush would not come out in favor of one over the other. Second, any new initiative for human exploration of the Moon

---

[30] Mark Albrecht interview, tape recording, Arlington, VA, 3 July 2003; "Quayle Puts Damper on Manned Mars/Moon Mission Prospects," *Defense Daily*, 5 April 1989, pp. 22-3; Quayle, *Standing Firm*, pp. 177-190.AQ4

and/or Mars would be a long-term commitment—so there would not be any huge budgetary impact. Finally, this program would be utilized as the central organizing principle for the entire civil space program—so that everything from the Space Station to space transportation to planetary programs would be structured around accomplishing this continuing objective.[31]

On 5 April, with Darman on board, Quayle raised the idea of announcing a robust exploration initiative with President Bush at their weekly lunch meeting. By the end of the discussion, the first of many on the topic that would occur over the coming months, President Bush gave the go-ahead to plan for a 20 July 1989 announcement. Quayle wrote in *Standing Firm* that the president wanted to use the anniversary "to make a major address on the space program, a speech that didn't just look back toward former glory but ahead to bold new achievement." Furthermore, he viewed this as an opportunity to challenge the long-held belief that only Democratic presidents (e.g., Kennedy and Johnson) had visionary approaches to space. Finally, President Bush, who had been criticized for a lack of vision, viewed this as a chance to answer his detractors by putting the weight of the White House behind a bold human exploration program. The eventual placement of human exploration of Mars on the government agenda had its genesis in this meeting.[32]

Over the coming weeks, the White House began aggressively taking steps to finalize its core space policy team. The previous month, Administrator Fletcher had formally resigned as NASA Administrator. After receiving his resignation, the administration began working to fill the position. During this search, the name of a former astronaut rose to the top—Rear Admiral Richard Harrison Truly. Vice President Quayle wrote in his memoirs, "Dick Truly was a friend of [White House Chief of Staff John] Sununu and became his candidate; I didn't have a good candidate of my own, and so I went along with Truly's selection."[33] Truly joined NASA in 1969 as an astronaut. Prior to that, he spent nearly a decade serving as a naval aviator, test pilot, and member of the USAF Manned Orbiting Laboratory program. After serving as CAPCOM for the Apollo-Soyuz Test Project (ASTP) and Skylab missions, Truly served on one of the astronaut crews that conducted the Space Shuttle *Enterprise* approach and landing test flights. He made two flights into space, the first in November 1981 as the pilot of Space Shuttle *Columbia* and the second in

---

[31] Quayle, *Standing Firm*, pp. 177-190; Albrecht interview.

[32] Ibid.; Frank J. Murray, "Putting Man on Mars May be Bush's Goal", *Washington Times*, 20 July 1989, sec. A1; Howard McCurdy, *The Decision to Send Humans Back to the Moon and on to Mars* (Washington, DC: NASA History Division, 1992), pp. 4-13; Quayle, *Standing Firm*, p. 181.

[33] Quayle, *Standing Firm*, p. 181.

August 1983 as commander of the Space Shuttle *Challenger*—which was the first night launch and landing during the shuttle program. After this mission, he left NASA for several years to head the newly formed Naval Space Command. In 1986, he returned to NASA as Associate Administrator for Space Flight—he was widely credited with guiding the shuttle program back to operational status after the 1986 *Challenger* accident. Vice President Quayle wrote in *Standing Firm*, "as often happens in Washington, Truly got the job by default: he didn't have the sort of negatives that might make news and sink the nomination." Mark Albrecht recalled that the decision to make Truly administrator was made "rapidly, without a great deal of serious discussion or assessment."[34] Over the subsequent three years, this hasty selection of a true NASA insider and devoted "Shuttle Hugger" would come back to haunt an administration that needed somebody heading the space agency who shared the Space Council's objectives. Although the White House thought it was completing a policy triumvirate (including Quayle and Albrecht) capable of transforming NASA, Truly saw his job as protecting the space agency from danger. As a result, it eventually became clear that he did not share the Space Council's reform goals.[35]

On 12 April, President Bush officially introduced Admiral Truly as his nominee to head NASA at a ceremony in the White House Roosevelt Room—attended by key members of Congress. The president opened by saying, "this marks the first time in its distinguished history that NASA will be led by a hero of its own making, an astronaut who had been to space, a man who has uniquely experienced NASA's tremendous teamwork and achievement." Bush acknowledged that because Truly was still an active duty flag-rank officer with the Navy, he would have to attain a congressional waiver for his appointment—the National Aeronautics and Space Act (NAS Act) of 1958 prohibited military officers from heading the civilian space agency. The president then handed the podium over to the nominee, who thanked Bush for showing confidence in his ability to guide the space agency. In conclusion, Truly stated that he looked forward to working with Vice President Quayle and the National Space Council.[36]

The *Washington Post* and *The New York Times* ran articles the next day suggesting that news of Truly's nomination was met with widespread approval on Capitol Hill.

---

[34] Albrecht interview.

[35] Press Release, White House Office of the Press Secretary, 12 April 1989, Bush Presidential Records, Bush Presidential Library; Quayle, *Standing Firm*, p. 181; Albrecht interview; Kathy Sawyer, "Bush Taps Truly to Head NASA: Former Astronaut Popular on Hill," *The Washington Post* (13 April 1989).

[36] *Public Papers of the Presidents of the United States*, 20 January 1989, *Remarks Announcing the Nomination of Richard Harrison Truly To Be Administrator of the National Aeronautics and Space Administration*, http://bushlibrary.tamu.edu/papers/ (accessed 18 May 2002).

The *Post* article explained that Truly had initially expressed doubts regarding the post, because he was concerned that it would necessitate the forfeiture of his military pension. Despite some congressional wariness related to placing a military officer in charge of NASA, it was reported that his military status was not expected to be a problem. The newspaper further reported that Democratic Congressman Norman Mineta of California, an opponent of military influence in the space agency, was nevertheless supporting the appointment due to Truly's unmatched depth of experience. House appropriation committee staffer Richard Malow remembered thinking Truly was a good choice. "Dick Truly always came across as being very honest and he was an outstanding head of the Shuttle office."[37] Senate staffer Stephen Kohashi recalled feeling that Admiral Truly was a competent individual, with a significant technical background and a sincere excitement for the space program. "I…recall [feeling relieved] that somebody solid and steady was assuming the helm," Kohashi said, "…although some [people], I suspect, would have preferred someone with a…more dynamic image."[38] The only true criticisms of Truly were that he was "too much the technocrat to be the combination of diplomat and salesman that his new job requires…[and that he is] no fan of developing commercial space enterprises."[39/40]

---

[37] Richard Malow interview via tape recording, Washington, DC, 25 October 2004.

[38] Kohashi interview.

[39] Kathy Sawyer, "Bush Taps Truly to Head NASA: Former Astronaut Popular on Hill," *The Washington Post*, 13 April 1989; Warren E. Leary, "Bush Chooses Former Astronaut to Head NASA, in a First," *The New York Times*, 13 April 1989.

[40] On 23 June, the U.S. Senate confirmed the nomination of Admiral Truly. Along with the confirmation vote, the Senate passed S. 1180, legislation that would allow him to retain his status, rank, and grade as a retired military officer and guaranteeing his retirement benefits from his Navy service after he retired from civilian life. The bill also provided that Admiral Truly, as the NASA Administrator, shall be "subject to no supervision, control, restriction, or prohibition (military or otherwise) other than would be operative…" if he were not a retired Navy officer. Six days later, OMB Director Darman sent a memorandum to President Bush recommending that the latter authorize the appointment of Admiral Truly and sign S. 1180—the next day James Cicconi of the Office of Personnel approved that recommendation and sent the bill to President Bush for his signature, which was affixed in the normal course of business before the deadline of 10 July 1989. [Memorandum, Richard Darman to President Bush, 29 June 1989, Bush Presidential Records, George Bush Presidential Library.]

On 20 April, at a White House ceremony in the Old Executive Office President Bush signed an Executive Order establishing the National Space Council.[41] He opened his formal remarks by telling the attendees that he was fulfilling the promise he had made the previous year to create the legislatively mandated council. With Vice President Quayle serving as Chair, Bush stated his belief that, "the Space Council will bring coherence and continuity and commitment to our efforts to explore, study, and develop space…." The president stated his belief that the space program was essential to sparking the American imagination, which inspired young people to enter the fields of science, math, and engineering—keys to ensuring the nation's competitiveness in the future. He concluded by saying he was signing the executive order with "one objective in mind: to keep America first in space…it's only a matter of time before the world salutes the first men and women on their way outward into the solar system. All of us want them to be Americans."[42,43] With

---

[41] On 15 March, the Office of the Vice President had submitted to OMB Director Richard Darman a proposed Executive Order for establishing the National Space Council. The draft executive order stated that the goal of the space council was to "provide a coordinated process for developing a national space program and for overseeing the implementation of national space policy and related activities…." The Vice President, who would act as the primary space policy advisor to the President, would chair the council. The remaining members of the body would be the: Secretary of State, Secretary of Defense, Secretary of Commerce, Secretary of Transportation, OMB Director, White House Chief of Staff, National Security Advisor, Director of Central Intelligence, and the NASA Administrator. Upon the request of the Vice President, the Chairman of the Joint Chiefs of Staff, the Presidential Science Advisor, and heads of other executive departments and agencies could also be called upon to participate in meetings. The executive order also provided for the creation of the Vice President's Space Policy Advisory Board. This committee would be composed of private citizens appointed to advise the Vice President on national space policy issues. Vice Presidential Counselor Diane Weinstein wrote Darman that "given the urgent need for the Council to begin exercising its critical responsibilities as soon as possible…[the] Vice President recommends that the President sign the enclosed proposed Executive Order establishing the National Space Council." A week later, Darman received a memorandum from Bonnie Newman, Assistant to the President for Management and Administration, concurring with the OMB Director's recommendation to provide funding to the National Space Council under the auspices of Public Law 100-440, which provided budget resources to the Executive Office of the President for "Unanticipated Needs." Darman forwarded the memorandum to President Bush recommending an allocation of $181,000, which would allow the Space Council to begin operations in fiscal year 1989. [Letter with an attached draft Executive Order, Diane Weinstein to Richard Darman, 15 March 1989, Bush Presidential Records, George Bush Presidential Library; Memorandum, J. Bonnie Newman to Richard G. Darman, 22 March 1989, Bush Presidential Records, George Bush Presidential Library; Memorandum, Richard G. Darman to President George Bush, 23 March 1989, Bush Presidential Records, George Bush Presidential Library.]

[42] *Public Papers of the Presidents of the United States*, 21 April 1989, *Remarks on Signing the Executive Order Establishing the National Space Council*, http://bushlibrary.tamu.edu/papers/ (accessed 18 May 2002).

[43] The actual executive order had been amended somewhat since its original submittal to OMB the previous month. There were two additions made to the full membership of the council—the Secretary

the Space Council officially created and a core set of policy entrepreneurs in place, the administration was organized to begin a concerted effort aimed at shaping an initiative for human exploration beyond Earth orbit.[44]

The Policy Stream: The Ad Hoc Working Group

In the six weeks following the creation of the Space Council, as the administration was concentrating on other more pressing policy matters, no major actions were taken with regard to the future course of the space program. By the end of May, however, there was a flurry of activity to generate a policy initiative in preparation for the Apollo 11 anniversary. On 25 May, Mark Albrecht called Admiral Truly to ask whether NASA could return to the Moon by the end of the century—in preparation for a Mars mission early in the next century. Albrecht was stunned by Truly's response. "His first reaction was 'don't do it.' NASA cannot handle this." The NASA Administrator was unsure whether this request was simply Albrecht playing 'what if' games, or whether this was a serious proposition. As a result, he called Vice President Quayle, who confirmed that both he and President Bush wanted to know whether this was something NASA could accomplish. After consulting with Frank Martin, Director of NASA's Office of Exploration, Truly concluded that there was

---

of the Treasury and the President's Science Advisor. The functions of the council were also fine tuned, and were listed in the final order as follows:
  A. The Council shall advise and assist the President on national space policy and strategy…
  B. The Council is directed to:
     1. Review United States Government space policy, including long-range goals, and develop a strategy for national space activities;
     2. Develop recommendations for the President on space policy and space-related issues;
     3. Monitor and coordinate implementation of the objectives of the President's national space policy by executive departments and agencies; and
     4. Foster close coordination, cooperation, and technology and information exchange among the civil, national security, and commercial space sectors…
  C. The creation and operation of the Council shall not interfere with existing lines of authority and responsibilities in the departments and agencies.
The rest of the document was substantively the same as the draft order—including provisions detailing: the responsibilities of the chairman, the national space policy planning process, the establishment of the Vice President's Space Policy Advisory Board, and the requirement to "submit an annual report setting forth its assessment of and recommendations for the space policy and strategy of the United States Government." [Executive Order Establishing the National Space Council, 21 April 1989, Bush Presidential Records, Bush Presidential Library.]

[44] One could argue that Admiral Truly was not actually a committed policy entrepreneur for human spaceflight beyond Earth orbit, and that in fact he worked against the program. Regardless, he became one of the key players in the SEI process and was heavily involved in assembling the administration plan and trying to sell it on Capitol Hill. It seems that this qualifies him as a policy entrepreneur for the purposes of this manuscript.

no way he could rebuff a presidential initiative. Albrecht recalled later "his initial impulse turned out to be quite revealing, because in the end, NASA couldn't handle it."[45] What is equally revealing, however, is the fact that nobody at the White House reconsidered the wisdom of announcing a new initiative given the agency's reluctance.

After this interaction, the Space Council staff concluded that it needed to get a better sense of the correct technical approach to get back to the Moon on a permanent basis and then on to Mars. To this end, Mark Albrecht set up a meeting for the end of the month with senior NASA leaders to discuss alternatives. On 31 May, Truly and Martin met to discuss a potential initiative. A few years later Martin recalled the discussion:

> The nature of that conversation …was that going to the Moon [was] not the right answer. We have been to the Moon. If we are going to go to the Moon, we need to go back to stay. In the process of doing that, if you announce that you are going to go to the Moon and then go to Mars with humans, you had better be prepared to send robots along in the process.

Although he had signaled to Albrecht just days before that announcing any initiative at all was unwise, Admiral Truly was now supporting a much more aggressive (not to mention expensive) long-term exploration strategy. Later in the day, Truly, Martin, NASA Deputy Administrator J.R. Thompson, and former NASA Associate Deputy Administrator Philip Culbertson met with Albrecht to discuss proposals for a potential initiative. At this meeting, Truly told Albrecht that he "believed that the real program was Earth, Moon, and Mars as a total program strategy with both man and machines working together. It is that program that I think we need to proceed with." Albrecht did not challenge the addition of Mars exploration to the initiative, even though his original inquiry had been limited to Moon exploration. By the end of the meeting, he had tasked NASA with preparing options and recommendations for a presidential decision to take advantage of the unique opportunity of July 20th and to achieve significant milestones by the end of the century.[46] Albrecht did not specifically ask NASA to consider the fiscal repercussions of a Moon-Mars initiative, although Admiral Truly made it clear this was not going to be cheap.[47]

---

[45] McCurdy, *The Decision to Send Humans Back to the Moon and on to Mars*, pp. 12-13; Albrecht interview; Frank Martin interview by Howard E. McCurdy, in *The Decision to Send Humans Back to the Moon and on to Mars* (Washington, DC: NASA History Division, 1992).

[46] Ibid.

[47] Douglas O'Handley interview via electronic mail, Morgan Hill, CA, 22 November 2004.

Chapter 3: Bush, Quayle, and SEI

With official direction from the White House, Admiral Truly moved quickly to establish a working group to pull together the alternatives for a Moon-Mars initiative. He immediately called NASA Johnson Space Center (JSC) Director Aaron Cohen and asked him to gather a group of experts to compile program concepts. Truly asked Cohen to keep this activity confidential in order to keep the space agency's efforts a secret. From a technical and programmatic perspective, Dr. Cohen was an excellent choice to lead this policy alternative generation process. He had joined the space program in 1962, serving as program manager for Project Apollo's Command and Service Module and the Space Shuttle Orbiter Project. Given his rich background with the human exploration program, Cohen was the logical choice to lead this effort. Furthermore, he was very enthusiastic about the initiative and believed it was a good thing for the entire country. He believed from the start, however, that considerable monetary resources would be required to successfully implement the program—not to mention a long-term commitment from the executive branch, the legislative branch, and the American public.[48]

Cohen assembled a small team that included Frank Martin, John Aaron, Mark Craig, Charles Darwin, Mike Duke, and Darrell Branscome—this became known as the Ad Hoc Working Group (AHWG).[49] On 4 June, just six weeks before the president was to deliver his address, the AHWG assembled at Johnson Space Center. Mark Craig, Director of the Lunar and Mars Exploration Office at JSC, remembered that because "Admiral Truly wanted to keep this extremely secret, for obvious reasons…Aaron Cohen found a building in the back lot of JSC that … was secure. So we set up headquarters back there. It already had computers in it. It was ready to move in. It was locked. So we basically set that up as our center of operations." The goal of the team was to pour over the available information from the National Commission on Space, Ride Report, and Office of Exploration. Within two weeks, the AHWG was to develop a set of briefing charts that scoped out, in terms of cost and schedule, what would be required to return humans to the Moon by the year 2000. Frank Martin recalled that there was no effort to make it cheap, although there was some discussion about the feasibility of the initiative in the current political environment. Admiral Truly actually expressed his opinion that it was more important to "Do it right. Make sure we can do this. Make sure we understand the scope and magnitude of this program."[50] This necessarily meant that the AHWG would not provide alternatives with different budget profiles, although Mark Albre-

---

[48] Aaron Cohen interview via electronic mail, College Station, Texas, 9 December 2004.

[49] Martin interview.

[50] Ibid.

cht had implicitly asked for multiple options. Instead, it would provide the Space Council with what it believed to be the right answer—regardless of cost. Although this didn't cause considerable friction at this early point in the process, this agency approach would eventually lead to an increasingly bitter relationship with Vice President Quayle and the Space Council staff.

The AHWG split up its work to create a long-term exploration strategy—Mark Craig led the technical analysis, John Aaron led the cost analysis, Darrel Branscome led the future planning analysis, Charles Darwin led the space transportation analysis, and Mike Duke led the science analysis. The vast majority of the work was done under Mark Craig's leadership, utilizing his staff within the Lunar and Mars Exploration Office. The AHWG met as a consulting body, working to shape the various inputs from these engineers into a briefing for the White House.[51] There was some concern within the agency regarding the planning monopoly that had emerged. Douglas O'Handley, the Deputy Director of the Office of Exploration and a veteran of the Jet Propulsion Laboratory (JPL) in California, remembered that the AHWG was a closed group composed almost exclusively of engineers from JSC. There was no effort to involve other NASA centers in this initial planning process. The environment within the agency during this period lacked the collegiality that had been experienced under the leadership of the Office of Exploration as agency-wide reports like "Beyond Earth Boundaries" were being drafted. O'Handley recalled that as the AHWG developed its plans, the JSC team clearly wanted as little JPL involvement as was possible. He became increasingly concerned as the planning moved forward because he felt "there was clearly a naiveté about the impact of life sciences on the whole initiative. From my background, I was beginning to see holes in the fabric and things that JSC didn't know much about falling to the side."[52]

On 13 June, Mark Craig presented the AHWG program concept to Admiral Truly and Associate Administrator for Spaceflight Bill Lenoir at a secret meeting held in Washington, DC. The briefing, entitled "A Scenario for Human Exploration of the Moon and Mars," proposed an approach that would start out with lunar activity and robotic missions—which would be precursors to Mars exploration. The AHWG approach required a sharp jump in the agency's yearly appropriation, with stable annual investments for the life of the initiative. It was believed that after

---

[51] Frank Martin interview by Howard E. McCurdy, in *The Decision to Send Humans Back to the Moon and on to Mars* (Washington, DC: NASA History Division, 1992); Mark Craig, interview by Howard E. McCurdy, in *The Decision to Send Humans Back to the Moon and on to Mars* (Washington, DC: NASA History Division, 1992).

[52] O'Handley interview.

successfully establishing a lunar base and completing robotic precursor missions, a funding wedge would open providing the resources for Mars exploration.[53] The plan called for three phases of lunar development: emplacement, consolidation, and utilization. During the emplacement phase, extending from 2000 to 2004, a lunar station consisting of a base camp and science outpost would be assembled to house a crew of four on six-month tours of duty. It was expected that this initial capability could be accomplished with only six Shuttle-Cargo (Shuttle-C)[54] launches and a single crewed Shuttle launch. During the consolidation phase, extending from 2003 to 2006, a constructible habitat would be added to the lunar base—raising the crew size to eight and lengthening tours of duty to one year. During the utilization phase, extending from 2006 to 2017, a lunar oxygen production capability would increase crew size to 12 and lengthen tours of duty to three years—providing the capacity for significant scientific work and certification of Mars exploration hardware. As astronauts constructed and prepared the lunar station to be operational, NASA would begin robotic precursor missions to the red planet. These missions would conduct high-resolution imaging of the planetary surface, long-range surface roving, and return samples to scientists on Earth. The AHWG believed that crewed missions to Mars could begin in the 2015 timeframe, with a crew of four reaching the planet (after a Venus flyby) for a 100-day nominal stay in the Mars system—50 days on the surface. The intent was that successive missions would reach the Martian system faster, with longer surface stays up to two years for crews of five.[55]

The AHWG scenario was founded on a number of fundamental ideas regarding available technologies and infrastructure. First, the entire approach was based upon the assumption that Space Station Freedom (SSF) would be utilized as the hub for assembly work to construct lunar and Martian transfer vehicles. This required that a Shuttle-C be developed, capable of launching 68 metric-ton payloads into Earth orbit. Second, the strategy would eventually require reusable lunar transfer vehicles (LTV) and lunar excursion vehicles (LEV)—capable of conducting five missions without major maintenance before mandatory replacement. The LTV would utilize aero-braking technology for return to SSF, and a lunar fuel production capability would be initiated for LEV return to lunar orbit. Third, Mars exploration spacecraft would depart from Earth orbit after being assembled by astronauts stationed at SSF.

---

[53] Ibid.; Mark Craig, "A Scenario For Human Exploration of the Moon and Mars," presented to Admiral Richard Truly on 13 June 1989.

[54] Shuttle-C was a 1980s proposal to use the Shuttle's infrastructure to create a heavy launch vehicle. This vehicle would have used the Space Shuttle's solid rocket boosters, external tank, and main engines. Instead of a crew-carrying orbiter, however, it would have used an expendable cargo carrier.

[55] Craig, "A Scenario For Human Exploration of the Moon and Mars."

Fourth, production of liquid oxygen (LOX) on the lunar surface would be required to open a significant cost wedge for Mars exploration. Douglas O'Handley remembers that at this point the budget estimates for the entire program ranged from a low of $85 billion to a high of $365 billion. The $85 billion estimate included a lot of risk, while $365 billion incorporated significant redundancy to reduce risk. It was felt within the agency that "these costs, compared to the defense budget for one year, seemed reasonable for a 20 to 30 year endeavor."[56] Although Bill Lenoir raised concerns regarding the necessary acceleration of space station construction to meet the objectives of the AHWG plan, the briefing was generally well received. No one expressed trepidation regarding the adoption of a program that would require a significant increase in the NASA budget, at a time when the federal government was in the midst of a serious fiscal crisis. Likewise, senior agency leaders did not question the complex and costly three-phase AHWG approach.

Two days later, the AHWG presented its proposal to Mark Albrecht. Mark Craig, who had never met Albrecht before, remembered being impressed with him. "The meeting...opened up with a monologue on why this was important and the problems that civilian space had had and was having, and that this was a way to fix them. I thought he was right on the money, having come from [the Space Station program, which] was suffering from a lack of definition of a strategic horizon." The briefing highlighted the AHWG approach, which was: lunar base, robotic exploration of Mars, and human exploration of Mars. It also included links to Mission to Planet Earth (MTPE), which Albrecht requested be removed from the briefing because he feared it would muddle the Moon-Mars focus. Later in the day, Admiral Truly, Frank Martin, Craig, and J.R. Thompson traveled across town to brief Vice President Quayle in his office in the Old Executive Office Building.[57] Frank Martin recalled later that Admiral Truly introduced the briefing by stating that Mars was "the long-term goal. It wasn't a program to go to Mars. It was a program to expand human presence [and] he talked about why it was important to do that." After Truly's introduction, Martin presented the primary elements of the AHWG approach. He was forthright with regard to the estimated cost of the exploration program, which had risen to $400 billion. This revised budget number was partially driven by an Administration request that crew safety be placed at 99.9999%, which meant that the probability of an accident occurring that resulted in a loss of crew was once every million flights. This high level of safety led to additional cost.[58] The AHWG plan would require increasing the space agency's budget by 10% annually until it

---

[56] O'Handley interview.

[57] Ibid.

[58] O'Handley interview.

reached $25 to $30 billion—doubling the current appropriation.[59] Truly concluded the briefing by saying that NASA could not fulfill this mission without an increased budget that would provide the resources to hire essential personnel and construct new facilities. Frank Martin later remembered that:

> [Quayle] was very interested. He was very friendly. He was wide-eyed and enthusiastic about it. He asked the kinds of questions you might expect to be asked from someone who is a non-technical type…the message I came away from that briefing at the White House [with] was the fact that for the first time in 20 years, somebody in the White House gave a damn about the Moon and Mars. That was what was very profound about it. He was willing to take the time and the effort to try to make something happen.

Overall, the NASA participants left the meeting with a very positive feeling that both Quayle and Albrecht would be willing to fight the necessary battles to make the exploration program work.[60]

The following day, Admiral Truly returned to the White House to meet with Chief of Staff John Sununu. Himself an engineer, Sununu moved through the briefing materials very quickly before signaling his support for the initiative. He told Truly that "investing in these kinds of things was good for the country and that he didn't care who made the content of the program. It was the fact that they were doing it that was important."[61] He said that he would leave the details to NASA, but that it wasn't feasible for the space agency to get a further 10% increase in the FY 1990 budget request. At the end of the meeting, Sununu made three requests. First, he wanted NASA to modify the program so that no money was required for the upcoming fiscal year. Second, he wanted the plan to be revised to present the President with options—not just lunar outpost, robotic Mars exploration, and Martian outpost. Many within NASA, most notably those outside JSC, thought that asking for additional alternatives "was absolutely the right thing to request."[62] Finally, before the President made his speech, Sununu wanted the benefit of having others outside NASA review the proposed exploration program.[63]

---

[59] This budget estimate had been calculated by OMB cost analyst Norine Noonan, in consultation with NASA.

[60] Martin interview; Craig interview.

[61] Martin interview.

[62] O'Handley interview.

[63] Martin interview.

Over the course of several weeks, Truly, Martin, Craig, and Darrell Branscome worked to refine the NASA proposal. They developed three different options for President Bush to consider:

- Lunar Outpost, then to Mars (NASA's recommended approach)
  - First crewed lunar landing in 2001 (crew of four, 30-day surface stay)
  - Expansion to 8 crew capacity by 2005
  - Expansion to 12 crew capacity by 2009 (1-year surface stay)
  - First crewed Martian landing in 2016
- Direct to Mars
  - First crewed Martian landing in 2008 (crew of four, 30-day surface stay)
  - Expansion to 8 crew capacity by 2014
  - Expansion to 12 crew capacity by 2018 (180-day surface stay)
- Robots Only

Due to the fact they could not expect any funding in FY 1990, the agency slipped the deadlines one year—so the Moon landing would not occur until 2001. Interestingly, based on NASA's analysis, the Direct to Mars option did not entail a significant cost reduction. This option would still require the NASA budget to increase to nearly $30 billion annually, with a peak of over $35 billion during the late 1990s. As one chart in the final briefing indicated, this would represent a larger federal investment (in real dollar terms) than Project Apollo and would raise the NASA share of the federal budget to 2.2%.[64] The agency did not provide any human exploration options that had significantly cheaper cost profiles.

The Political Stream: Briefing Key Actors

The revision process continued until just before Independence Day, after which the White House had arranged briefings for outside interest groups. For three days starting on 5 July, the Administration undertook a series of briefings to explain the Civil Space Exploration Initiative to four groups from outside the administration. The first group, which the White House labeled "Space Advocates," was composed of influential members of the space policy community not affiliated with a particular government agency or private sector company. This group included former Apollo 11 astronaut Mike Collins, Cal Tech professor Bruce Murray, former NASA Administrator Tom Paine, and President of the Planetary Society Louis Friedman.

---

[64] NASA, "Civil Space Exploration Initiative," presented to Vice President Dan Quayle on 15 June 1989.

Chapter 3: Bush, Quayle, and SEI

The briefing, which was conducted in the Indian Treaty Room of the OEOB, was very well received. Mark Albrecht recalled that the group was "obviously excited about it, very enthusiastic." Tom Paine, who had chaired the National Commission on Space, was extremely supportive and stated that this was exactly the kind of strategic direction that the American space program needed. There were universal strong positive statements; the only thing that the advocates questioned was the appropriate balance between Moon and Mars exploration. Both Mike Collins and Bruce Murray had previously come out in favor of a direct to Mars approach, so they were a little uncomfortable with NASA's recommendation to start with a return to the Moon. The rest of the group was largely in favor of NASA's strategy.[65]

The second group was composed of representatives from the science community. Frank Martin later remembered that this group was very supportive, "they were enthusiastic about it more than I would have imagined. They [agreed that] this is the right thing. And doing [the] Moon and then Mars is the right way to do it. It was pretty universal."[66] Surprisingly, no one from this group made a strong argument in favor of solely robotic exploration, perhaps sensing that Vice President Quayle was strongly in favor of the Moon-Mars approach. The third group was composed of chief executive officers from major U.S. corporations—many who were important NASA contractors. Mark Albrecht recalled that "industry was excited…but they were nervous about what [existing] programs could get cut to fund it. Anytime you hit the reset button in Washington, you find that everyone gets very nervous." Mark Craig remembered this being the most disappointing of the meetings because the level of industry support was not as robust as had been anticipated. The general reaction was that if the government wanted to do this, and was willing to put up the funding, then industry would get on board.[67] Douglas O'Handley recalled that the CEO's were concerned that the U.S. did not have the technical manpower to carry SEI off, despite the fact the administration believed the initiative would promote science and engineering education.[68] As a result, there were no impassioned speeches arguing that this was exactly the kind of bold long-term plan that the aerospace industry and national economy needed. Craig stated later, "I felt the Vice President was…knocked on his heels. He tried to elicit some kind of emotion and response from these people."[69]

---

[65] Ibid.; Craig interview; Albrecht interview.

[66] Martin interview.

[67] Craig interview.

[68] O'Handley interview.

[69] Ibid.

The final group, made up of key Congressional staffers, was by far the most dynamic. Mark Albrecht contends that during the Reagan administration the House "Appropriations Committee and the Appropriations staffers essentially ran the space program because NASA got no direction or interest out of the White House...the vacuum was filled by the appropriators." As a result, this was by far the most skeptical group—they were doubtful about the White House taking a renewed interest in space policy making and were not convinced that selecting an expensive new initiative was the best approach for providing the space program with direction. Led by Richard Malow, this group was most concerned about the potential budgetary impact of such a large undertaking. Malow was the most powerful staffer on the House Appropriations Subcommittee that funded NASA. Senator William Proxmire of Wisconsin had recently called him the space agency's 'shadow administrator' because he had so much influence over national space policy.[70] Malow had been working on space policy issues since 1972, far longer than NASA's senior leadership, and was well-known for pushing the space agency to emphasize affordable space science missions rather than expensive human exploration programs. In fact, a week earlier the *Wall Street Journal* had run a front-page article stating that while Malow "would love the U.S. to mount an expedition to the far side of the Moon and build a telescope there...such dreams are 'moot' because of the budget crunch. Instead...NASA [should] focus on more attainable goals."[71]

At the White House briefing, Malow remembered his "...initial reaction was that maybe this is something that we ought to be doing, but I don't think I jumped in and said 'that's the greatest idea in the world.' And as I started to see the details of it, as they unfolded, I became concerned, especially given the budget situation."[72] The reactions of other staffers were much more animated. For example, Stephen Kohashi, an aide to Senator Jake Garn, asked the briefers, "have you lost your mind?"[73] Kohashi said later, "politics is the art of the possible, and so it is with budgetary politics. I recall being incredulous at the magnitude of the price tag [for] the proposed program...and feared that it would have no credibility or viability on Capitol Hill."[74] In the end, no real champions emerged on the Congressional side. Instead, the meeting served to generate "a certain tension...between the Space

---

[70] Bob Davis, "Quiet Clout: How a House Staffer Wields Great Power Over Policy Decisions," *The Wall Street Journal* (30 June 1989), p. 1.

[71] Ibid.

[72] Malow interview.

[73] Martin interview.

[74] Kohashi interview.

Council staff and the staffs of the various committees on the Hill."[75] Malow recalled that one reason for this rising animus was the failure of the administration to consult with Congressional leaders as it was formulating a plan for the new human exploration program. Malow stated that such discussions would have led to "warnings about the overall budget situation, which is what we were concerned about. We may have tried to convince them that they needed to think it through a little bit more."[76] Despite this unenthusiastic response from the Congressional staffers, who would ultimately have a great deal of influence regarding the actual adoption and implementation of SEI, the White House marched forward with its plans to announce the initiative.

Starting on 13 July, just a week before the president's planned speech, Vice President Quayle and Admiral Truly began meeting with key members of Congress. These meetings were intended primarily to acquaint the legislators with the initiative before the President announced it nationally. That morning, the two briefed a group of Representatives and Senators with responsibility for space policy at a breakfast meeting. As with the Congressional staffers, the reactions were not wholly positive. In particular, House Appropriation Committee Chair Bob Traxler of Michigan wondered where the Administration was "coming from, we can't afford this…we've got other things on our plate, outside NASA."[77] A few days later, Quayle and Truly went to Capitol Hill to personally brief Senator Barbara Mikulski of Maryland, Chair of the Appropriations Subcommittee overseeing NASA's budget—who had been unable to attend the breakfast meeting. Frank Martin remembered later, "she was very supportive. She said "the budgets are going to be tight [but] I am glad the Administration is finally taking an interest in space." One final briefing was given to Senator Ernest Hollings of South Carolina, chair of Commerce, Science & Technology committee—once again, the White House received support for the program.[78] Regardless of the general support that the administration received from Mikulski and Hollings, the clear skepticism of Traxler was more important. As chair of the House appropriations committee, he would have enormous influence over the actual adoption of this program. Therefore, even before it was announced, SEI faced a huge uphill battle to gain backing from Congress. As Malow indicated, this was at least partially because the White House did not consult with Capitol Hill during the formulation of the new plan. More important, however, was the fact that there were no great supporters for implementing an expensive new program given the fiscal crisis facing the nation.

---

[75] Martin interview.

[76] Malow interview.

[77] Ibid.

[78] Martin interview.

Joining the Streams:
Human Exploration of Mars Reaches the Government Agenda

As these events were unfolding in Washington, President Bush was in Europe on a 10-day trip that included an address before the Polish National Assembly, a meeting with Solidarity leader Lech Walesa, a meeting with Hungarian leaders, and attendance at a G-7 summit in France. While he was away, Bush had essentially delegated decision-making responsibility for the exploration initiative to Vice President Quayle. Over the course of the previous month, Bush had discussed the development of the exploration initiative with Quayle at several of their weekly lunch meetings, but the president had essentially let his vice president make all the critical decisions with regard to the strategic plan. One important facet of their discussions was whether the Administration should set a target date of 2010 for completion of a Moon base and 2020 for an expedition to Mars. Although this debate continued up until the last moment, the two ultimately decided against specific deadlines because they feared it would adversely impact future budget deliberations. By early July, the President had fully committed to the program.[79]

On 14 July, Quayle chaired a meeting of the full Space Council to discuss the forthcoming announcement of the exploration initiative. Mark Albrecht recalled that "everyone lined up, thought it was a great idea and made a recommendation to the President that he go ahead and do this." Thus, when Bush arrived back in Washington two days before the speech, everything was already in place for him to announce the new plan.[80] Vice President Quayle wrote in his memoirs that if the agenda setting process for SEI sounded like a "…somewhat ad hoc, improvisational way to think about going to Mars, you're right. But what was important right then was to think big, to put a bit of 'the vision thing' back into the program, to get people excited about it once again, even if that meant getting ahead of ourselves." Quayle believed the only thing that would enliven the American people was a restoration of wonder in the idea of sending people to explore space, not just orbit around the Earth.[81]

Before the new initiative was officially announced, the 17 July 1989 edition of *Aviation Week and Space Technology (AW&ST)* broke the story that a secret White House review was considering a human lunar base and Mars initiative. The article opened by stating, "A sharp debate has been sparked within the Bush Administra-

---

[79] Albrecht interview; Karen Hosler, "Bush Unveils Moon, Mars Plans But Withholds Specifics," *The Baltimore Sun,* 21 July 1989.

[80] Ibid.

[81] Quayle, *Standing Firm,* pp. 177-190.

Chapter 3: Bush, Quayle, and SEI

tion and Congress by Vice President Dan Quayle's proposal that President Bush commit the U.S. this week to developing a manned lunar base as a stepping-stone to a manned flight to Mars. Under the proposal, the U.S. could build a lunar outpost by 2000-2010 and use the experience gained on the moon to develop that capability to mount a manned Mars mission by 2020." The magazine reported that Quayle had been formulating the initiative in secret meetings with a group of NASA officials, Mark Albrecht, and White House Chief of Staff John Sununu. Administration officials were quoted as saying that President Bush would not make a Kennedy-style call for reaching Mars within a specific timeframe, instead endorsing "the lunar base and manned Mars concepts as overall 21st century goals [and deferring] specific program and budget decisions on these goals until NASA completes a more intensive assessment of the mission options." The magazine reported that NASA's budget would have to double within a decade to pay for the initiative. This was at the same time that the House Appropriations Committee was planning on cutting NASA's FY 1990 appropriation by more than $1 billion, including a 50% decrease in funding for technologies key to Mars exploration. While Vice President Quayle recognized that the federal government faced serious budgetary limitations, he was quoted as saying that "when we have tight budgets, there will be winners and losers, but I am convinced a winner will be space." Craig Covault of *AW&ST* reported that NASA leaders saw a presidential endorsement as an opportunity to seek increased funding and begin serious mission planning. Overall, the article was uncannily accurate and set the stage for President Bush's upcoming address.[82]

On Thursday, 20 July 1989, with the decision in favor of an aggressive program for human exploration of the Moon and Mars made, President Bush prepared to announce the initiative at the anniversary celebration of Apollo 11's landing on the Moon twenty years earlier. At shortly before 10:00 a.m., President and Mrs. Bush, accompanied by Vice President and Mrs. Quayle, departed the White House for the short drive across the National Mall to the Smithsonian's National Air and Space Museum. Upon their arrival at the museum, the group was escorted to the Lunar Module display, where they were greeted by Admiral Truly, Neil Armstrong, Michael Collins, Buzz Aldrin, and Secretary of the Smithsonian Institution Robert Adams. After a quick photo opportunity attended only by invited pool photographers, President Bush and his growing entourage were escorted to the museum's front steps, where after a brief hold he was ushered on stage with an obligatory rendition of "Hail to the Chief."

---

[82] Craig Covault, "Manned Lunar Base, Mars Initiative Raised in Secret White House Review," *Aviation Week & Space Technology* (17 July 1989), pp. 24-26; Michael Mecham, "House Panel Proposes $1-Billion Cut for NASA," *Aviation Week & Space Technology* (17 July 1989), p. 26.

Mars Wars

President Bush, Vice President Quayle, and the Apollo 11 crew (NASA Image 89-H-382)

The first order of business for the event was the unveiling of an Apollo 11 postage stamp by Postmaster General Anthony Franks. The $2.40 stamp depicted Neil Armstrong and Buzz Aldrin raising the American flag on the plains of the Sea of Tranquility. After brief remarks by Truly, Armstrong, Collins, and Aldrin, Vice President Quayle introduced George Bush. President Bush opened his remarks by saluting "three of the greatest heroes of this or any other century: the crew of Apollo 11." Bush used the first several minutes of his address remembering the remarkable accomplishment of that first human landing on the lunar surface. He recounted his family's personal recollections of the landing—his children spread throughout North America, each listened in their own way. "Within one lifetime," the president stated, "the human race traveled from the dunes of Kitty Hawk to the dust of another world. Apollo is a monument to our nation's unparalleled ability to respond swiftly and successfully to a clearly stated challenge and to America's willingness to take great risks for great rewards. We had a challenge. We set a goal. And we achieved it."

Chapter 3: Bush, Quayle, and SEI

President Bush and Postmaster General Anthony Frank unveil Apollo 11 commemorative stamp (NASA Image 89-HC-394)

Celebrating such an important legacy, Bush asserted, was an appropriate time to look to the future of the American space program. He proclaimed the inevitability of human exploration and permanent settlement of the solar system in the 21st century, in the process confirming the United States's place as the preeminent space faring nation on Earth. Based on this rhetorical foundation, Bush unveiled his vision for this future exploration and settlement. "In 1961 it took a crisis—the space race—to speed things up. Today we don't have a crisis; we have an opportunity. To seize this opportunity, I'm not proposing a 10-year plan like Apollo; I'm proposing a long-range, continuing commitment. First, for the coming decade, for the 1990s: Space Station Freedom, our critical next step in all our space endeavors. And next, for the new century: back to the Moon; back to the future. And this time, back to stay. And then a journey into tomorrow, a journey to another planet: a manned mission to Mars." The President stated these missions would follow one another in a logical progression, creating a pathway to the stars. He made clear that while setting the nation on this visionary course, the primary focus of his Administration would be the completion of Space Station Freedom—a crucial stepping stone for missions beyond Earth orbit.

Mars Wars

President Bush signs Space Exploration Day proclamation (NASA Image 89-HC-402).

President Bush announced that he was tasking Vice President Quayle to "lead the National Space Council in determining specifically what's needed for the next round of exploration: the necessary money, manpower, and materials; the feasibility of international cooperation; and develop realistic timetables—milestones—along the way." He requested that the Space Council report its findings to him as soon as possible, with concrete recommendations regarding the proper course to the Moon, Mars, and beyond. As his remarks wound down, Bush explained the one rationale for the grand initiative by alluding to the Apollo 1 fire and the *Challenger* accident, stating, "there are many reasons to explore the universe, but ten very special reasons why America must never stop seeking distant frontiers; the ten courageous astronauts who made the ultimate sacrifice to further the cause of space exploration. They have taken their place in the heavens so that America can take its place in the stars. Like them, and like Columbus, we dream of distant shores we've not yet seen. Why the Moon? Why Mars? Because it is humanity's destiny to strive, to seek, to find. And because it is America's destiny to lead." The President opined that humans would ultimately reach out to the stars and to new worlds. While he believed that this would not happen in his lifetime or that of his children, making this dream a reality for future generations must begin with a commitment by his generation. He concluded that "we cannot take the next giant leap for mankind tomorrow unless we take a single step today."[83]

---

[83] *Public Papers of the Presidents of the United States,* 20 July 1989, *Remarks on the 20th Anniversary of the Apollo 11 Moon Landing*; http://bushlibrary.tamu.edu/papers/ (accessed 18 May 2002).

Chapter 3: Bush, Quayle, and SEI

President Bush announces SEI on steps of National Air and Space Museum (NASA Image 89-H-380).

Shortly after President Bush finished his remarks, Admiral Truly was introduced by Press Secretary Marlin Fitzwater in the White House Briefing Room to answer questions regarding the President's speech. Truly's answer to the very first question of the press conference was surprising, considering he had been intimately involved with the decision-making process for SEI. Asked if there was a proposed date for the first human landing on the red planet, he replied, "no…I just, frankly, learned this morning what [President Bush's] direction was."[84] Following this rocky start, Truly stumbled through a series of questions regarding the specifics of the plan and the political practicality of obtaining Congressional support for such an ambitious undertaking. Asked whether the potential budget for the Moon-base portion of the President's plan would top $100 billion, he replied somewhat lamely that it would be affordable over the long-term. When pressed on the probable cost of the endeavor, Truly admitted that "we don't have any detailed NASA figures. We have, obviously, in the last several weeks, looked in gross terms at what it would cost,

---

[84] Press Briefing, Admiral Richard H. Truly, 20 July 1989, Bush Presidential Records, George Bush Presidential Library.

73

but there was no specific timetable and I have not presented the President with a specific and detailed list of budgetary requirements."[85] The press conference continued along this shaky path with a question regarding the timetable for announcing a specific plan and budget for the initiative. Truly was once again unable (or unwilling) to provide a specific answer to this question, vaguely answering that it would take a number of months. He rallied in the end with his answer to a question regarding the necessity to bring in foreign partners, stating, "I think we can afford to go it alone, although I think that's probably in the long run not what's going to happen. The world has changed since the 1960s in space. It's premature…to know where we're heading, but I would think [SEI will] have an international flavor."[86] In retrospect, what is most striking about this press briefing was the lack of specifics regarding the Administration's plans to gain Congressional approval for SEI. Rapid decision- making was required to formulate the initiative in time to announce it on the Apollo 11 anniversary. Consequently, the White House did not have the time to formulate a strategy for winning support on Capitol Hill. Likewise, the Space Council had not drafted a top-level policy directive to guide administration activities aimed at further defining the initiative. In the coming months, these shortcomings would derail SEI.

As Admiral Truly's briefing was ongoing in the pressroom, guests began assembling on the White House South Lawn for a celebration of Apollo 11's landing on the Moon. With picnic tables spread throughout the center of the lawn and a U.S. Navy band playing in the background, the guests sat down to partake of a lunch that included barbecue pork ribs, barbecue chicken, potato salad, and deep dish apple Betty with ice cream. Among the 300 distinguished attendees were 23 Apollo astronauts, 26 key members of Congress, and dozens of NASA officials. President and Mrs. Bush arrived at noon and were seated at a table near the bandshell with a group of special guests.[87]

After lunch, President Bush walked to the stage to deliver some brief remarks to the gathered celebrants. He warmed the crowd up with a little astronaut humor, joking that planning the barbecue was hectic because he was unsure whether they preferred their food grilled or in a tube. He continued to say that "as you might

---

[85] Ibid.

[86] Ibid.

[87] Residence Event Task Sheet, Barbecue to Commemorate the 20th Anniversary of the Landing on the Moon, 7 July 1989, Bush Presidential Records, George Bush Presidential Library; Menu, Barbecue Lunch: 20th Anniversary of the First Moon Walk, 20 July 1989, Bush Presidential Records, George Bush Presidential Library.

Chapter 3: Bush, Quayle, and SEI

White House picnic celebrating Apollo 11 anniversary (NASA Image 89-H-396).

expect from a former Navy pilot who lived much of his adult life in Houston, I, too, am a longtime supporter of the space program." As an example of this support, he pointed to the fact that the single largest percentage increase for any agency in his Administration's first budget proposal was for NASA. He told those assembled, "My commitment today to forge ahead with a sustained, manned exploration program, mission by mission—the space station, the Moon, Mars, and beyond—is a continuing commitment to ask new questions, to seek new answers, both in the heavens and on Earth. James Michener was right when he told Congress: 'There are moments in history when challenges occur of such a compelling nature that to miss them is to miss the whole meaning of an epoch. Space is such a challenge,' he said. Well, today's announcement is our recognition that the challenge was not merely one that belonged in the sixties; it's one that will occupy Americans for generations to come ... the American people, I'm convinced, want us back in space—and this time, back in space to stay." Bush concluded by stating that he looked forward to the day when a future president addressed, in similar fashion, the first Americans to walk on Mars, "now only children, perhaps your children."[88]

---

[88] *Public Papers of the Presidents of the United States*, 20 July 1989, *Remarks at a White House Barbecue on the 20th Anniversary of the Apollo 11 Moon Landing*; http://bushlibrary.tamu.edu/papers/ (accessed 6 June 2002.)

4

The 90-Day Study

"We are going to return to the Moon and journey to Mars because we must, because the United States needs to challenge itself in order to be ready for the new world of the 21st century, now just over ten years away."

NASA Administrator Richard Truly, 26 October 1989

The public reaction to President Bush's announcement of SEI was swift, and not altogether positive. The following day, the headline on the front page of *The New York Times* read, "President Calls for Mars Mission and a Moon Base: Critics Cite High Costs—Bush Offers No Timetable or Budget for Plan, Leaving That to Space Council." The article stated the speech set the stage for "the first full-scale debate in years on the nation's troubled space program." The piece cited expert opinions predicting the initiative would cost at least $100 billion, and could rise to as much as $400 billion.[1] The reaction from the Democrat-controlled Congress was largely critical. Senator Al Gore of Tennessee, Chairman of the Subcommittee on Space, Science, and Technology, stated that "by proposing a return to the Moon and a manned base on Mars, with no money, no timetable, and no plan, President Bush offers the country not a challenge to inspire us, but a daydream." His fellow senator from Tennessee, James Sasser, concurred, stating "the President took one giant leap for starry-eyed political rhetoric, and not even a small step for fiscal responsibility. The hard fact is this administration doesn't even have its space priorities established for next year, much less for the next century. We have numerous space and science-

---

[1] Bernard Weinraub, "President Call for Mars Mission and a Moon Base," *The New York Times*, 21 July 1989, sec. A1.

related programs already on the table, all of them worthy, all of them high-ticket and all of them competing for scarce dollars."[2] House Budget Committee Chairman Leon Panetta was quoted saying, "The budget deficit is stealing the resources we need to…resume our nation's mission in space. When this President is ready to recognize that we can't do all he would like to do even on this planet without new revenues, then perhaps we can talk about Mars."[3] Even fellow Republicans were wary. Representative Bill Green of New York, the ranking minority member on the subcommittee with oversight of the NASA budget, stated that "given the federal budget deficit and earthly demands, I don't see how we can afford expensive manned programs in space in the near future."[4]

The Baltimore Sun captured the mood very well, writing that the announcement of a human mission to Mars "was tempered by the financial worries that took much of the thrill out of America's romance with outer space after the historic flight of Apollo 11."[5] For this very reason, the American public was not terribly supportive of the new initiative. A Gallup Poll released shortly after the announcement suggested that only 27% of Americans believed space spending should be increased, and only 51% thought being the first nation to land a human on Mars was a meaningful goal.[6] Not surprisingly, *The Wall Street Journal* reported support from the aerospace industry for SEI, although many executives believed a strong lobbying effort would be required to get Congressional approval for the expensive undertaking. In a prepared response, Martin Marietta Chairman Norman Augustine stated, "we applaud the president's call for renewed vigor in pursuit of the space frontier and the many benefits it implies." Despite similar supportive statements flowing from other aerospace giants, there remained a sense of pessimism within most corners of the industry—the result of looming questions regarding the source of the billions of dollars needed to carry out the ambitious plan. "It's not the same clarion call that President Kennedy gave when he set a moon-landing [goal]," said aerospace analyst Wolfgang Demisch.[7]

---

[2] Ibid.

[3] David C. Morrison, "To Shoot the Moon, and Mars Beyond," *Government Executive* (September 1989), pp. 12-22.

[4] Weinraub, "President Call for Mars Mission and a Moon Base."

[5] Ibid.; Karen Hosler, "Bush Unveils Moon, Mars Plans But Withholds Specifics," *The Baltimore Sun*, 21 July 1989.

[6] Morrison, "To Shoot the Moon, and Mars Beyond."

[7] Roy Harris Jr., "Firms Rejoice Over Reborn U.S. Space Program," *The Wall Street Journal*, 24 July 1989.

Chapter 4: The 90-Day Study

There were somewhat diverse opinions regarding President Bush's speech amongst NASA's senior leaders. On one hand, many were impressed that Bush was able to depart from the written text during his speech, which they felt proved that Vice President Quayle really understood NASA's plan and had effectively briefed the president. Douglas O'Handley recalls that this was probably the most positive thing ever said about Quayle by NASA officials.[8] On the other hand, some agency officials believed the address laid the foundation for a ruinous relationship between the Space Council and NASA. Aaron Cohen felt this was the case because neither organization was paying attention to what the other was saying. Top NASA leaders thought the speech was a Kennedyesque declaration calling for a large-scale national effort to build a lunar base and send humans to Mars. Quayle and Albrecht, in contrast, wanted to introduce a new way of doing business within the space program—one that involved smaller budgets and more aggressive technology development (which they believed would lead to large cost efficiencies). Cohen recalled, "the day that the initiative was announced was a day of great elation [at NASA]. It took everyone back to the days of Apollo."[9] For a White House that wanted to change the Apollo paradigm, this was not the desired reaction from the space agency.

After the euphoria of the announcement died down, Douglas O'Handley argues "Frank Martin and Admiral Truly realized that they needed to back up the skeletal AHWG studies."[10] This effort would include validating data that had been presented to the Space Council and assessing the technology readiness levels for the equipment needed to carry out the initiative—this review became known as the 90-Day Study. Mark Albrecht remembered later that NASA "stepped forward and almost demanded to lead this effort, which indicated the beginnings of a little friction" between the space agency and the Space Council staff. From the NASA perspective, however, the study was initiated because President Bush's speech had provided the agency with a charter to develop a plan for a lunar base and human mission to Mars. There was a fundamental belief among senior leaders at the agency that this was exactly what the Space Council wanted. In fact, some of Albrecht's public statements at the time seemed to indicate this was the case. He was quoted in *Government Executive* magazine saying that now that a national space policy goal had been set, the council would "leave it to the departments and agencies to decide how they're going to achieve that. Once they've made that determination, we'll review that to see whether or not it comports with national policy, or whether it's

---

[8] O'Handley interview.

[9] Cohen interview.

[10] O'Handley interview.

realistic or plausible. But in terms of getting into their programs and plans for the purposes of 'We know a better way,' that's just not what we're here to do."[11] Later in the process, however, the White House had changed its tune, and both Albrecht and Vice President Quayle were arguing that they never wanted NASA to conduct the 90-Day Study. Aaron Cohen argues this problem arose because the Council didn't have its own ideas regarding how to start the process. Although it had announced the initiative, the administration didn't have a good sense regarding how to proceed after the speech. He contends this was the main reason problems emerged between the two organizations.[12]

In the end, Albrecht asked NASA to provide "a variety of different approaches ... we want a variety of time frames, we want a variety of cost profiles, we want a variety of technologies, so the President can choose among different options rather than being told 'this is how to do it.'"[13] This was not the method, however, that Admiral Truly favored. The Administrator simply wanted to pull together the wealth of data that had been generated during the preceding five years and draft a report that would be ready within three months.[14] Over the coming months, Truly was warned during two meetings of the full Space Council that his plan "was not the approach most members wanted to pursue."[15] The Council members wanted NASA to develop options based on innovative new technologies that could potentially offer reduced long-term costs.[16] Admiral Truly essentially disregarded this direction from the Council. At the same time, by allowing the space agency to pursue its own course, the Council in effect delegated the authority granted to it by President Bush to conduct a review of options for implementing SEI.

A week after President Bush's speech, Admiral Truly assigned Aaron Cohen to lead an agency-wide effort to fashion a plan for establishing a lunar base and exploring Mars, drawing upon existing NASA planning documents. During the remainder of the year, the space agency never wavered from this approach.[17] Although he was asked to examine both technical and management issues, Cohen chose to

---

[11] Morrison, "To Shoot the Moon, and Mars Beyond."

[12] Albrecht interview; Cohen interview.

[13] Albrecht interview.

[14] Martin interview.

[15] James Fisher and Andrew Lawler, "NASA, Space Council Split Over Moon-Mars Report," *Space News* (11 December 1989), p. 10.

[16] Ibid.

[17] Press Release 89-126, National Aeronautics and Space Administration, 27 July 1989, Bush Presidential Records, George Bush Presidential Library.

ignore questions regarding changes in NASA's management culture. Some believed this decision was fueled by his view that the new initiative was laying out another Apollo program—in essence, a reinstitution of Kennedy's mandate."[18] As a result, there was no need to change the management culture that had successfully landed humans on the Moon. The only task was to define an aggressive program to meet President Bush's new mandate. In his book *Dragonfly: NASA and the Crisis Aboard Mir*, author Bryan Burrough detailed a revealing conversation between Albrecht and Cohen after the latter had been named to lead the 90-Day Study:

"Most of all we want alternatives, plenty of alternatives," Albrecht told Cohen.

"What do you mean, alternatives?" Cohen asked. From the blank look on the JSC director's face, Albrecht could tell he wasn't getting through.

"Alternatives," Albrecht repeated. "I mean, there has to be more than one way to do this. Give us a Cadillac option, then give us the El Cheapo alternative, with the incumbent risks. Talk about all the different technologies that could be learned."[19]

Albrecht believed this interaction, and NASA's reaction over the coming months, was the beginning of a never healed rift between the Space Council staff and NASA.[20] Cohen later recalled the conversation differently. Although he remembered Albrecht asking for several options, he has no recollection of being asked to provide alternatives with significantly different cost profiles. Without this direction, mission planners at JSC felt the only course of action was to develop a program plan based on the President's speech.[21]

Frank Martin believed the cause of this burgeoning conflict was the differing approaches of the two organizations. The Space Council staff, with backgrounds largely in the national security space sector, wanted NASA to develop alternatives starting "with a clean sheet of paper." The JSC view was that it would be a shame not to take advantage of the research that had been conducted during recent years. In

---

[18] Wendell Mendell interview via electronic-mail, 15 September 2003.

[19] Bryan Burrough, *Dragonfly: NASA and the Crisis Aboard Mir* (New York, NY: HarperCollins Publishers, 1998), p. 240.

[20] Albrecht interview.

[21] Cohen interview.

fact, Mark Craig remembered later that "there were never any debates about using a 'clean sheet.' Our goal was to find the best approach to meet a set of requirements, not to just find something new for its own sake."[22] In the end, NASA employed the JSC methodology and began developing an SEI strategy that was highly dependent on past studies and didn't consider multiple alternatives with different budgetary requirements.[23] Douglas O'Handley contends, "this is where the initiative fell apart, when it was taken over by the Johnson Space Center."[24]

Waiting for NASA

By late August, the White House was getting gradually more worried about the progress NASA was making on the 90-Day Study. Mark Albrecht was concerned with the weekly status reports he was receiving from the Technical Study Group (TSG), which was the JSC-led team tasked with carrying out the study. "We didn't like the reaction we got from NASA," he remembered. "It had an 'uh oh' quality to it. NASA reports seemed to be full of lofty verbiage but few technical outlines or alternatives for what a lunar base and a Mars mission would actually look like."[25] Throughout this period, Albrecht kept emphasizing that the President wanted to see a lot of technical and budgetary options. Based on the space agency's responses, however, the council staff was beginning to get the strong feeling that it wasn't going to get any alternatives. Although Congress wasn't heavily engaged during this period, there was rising concern because of the increasingly frayed Space Council-NASA relationship. The feeling on Capitol Hill was that this strain was caused largely because NASA was "running their own plan, which wasn't the same as the White House's plan."[26]

As time went on, these stressed relations escalated into an all out war between the TSG and the Space Council. NASA's Douglas O'Handley had actually made a few friends among the Space Council staff, and they were pleading with him to provide assistance. In the end, however, he was not able to provide any support because Admiral Truly and the TSG controlled all information relating to the 90-Day Study. Things got so bad that every time senior NASA officials returned from a White House meeting, there was another story about "those dumb [expletive] on the Space

---

[22] Mark Craig interview via electronic-mail, 12 September 2003.

[23] Martin interview.

[24] O'Handley interview.

[25] Bryan Burrough, *Dragonfly*, p. 240.

[26] Malow interview.

Council. I have often thought," O'Handley stated later, that the conflicting "personalities caused many of the problems. If, instead of fighting with the Space Council, we had tried to work with them, the outcome might have been different."[27]

While this external battle was being waged between the Space Council and NASA, there was another internal battle being waged within the agency. There was rising apprehension regarding JSC's control of strategic planning for the initiative. Although the TSG was to a degree soliciting advice from other field centers, there was a feeling that the JSC leadership didn't really take outside advice very well. Douglas O'Handley argued later, "I absolutely think a wider net should have been cast within NASA, but JSC deprived the other centers an opportunity to contribute to the initiative."[28] The aerospace industry also wanted to play a role in the mission development, but weren't heavily involved. Although there were numerous technical concepts and architectural options floating about, the TSG essentially ignored them. JSC became "Fortress NASA" and outside ideas were not welcome.[29]

Despite ongoing problems between the Space Council and NASA, and misgivings about the initiative on Capitol Hill, the TSG was allowed to continue compiling the 90-Day Study. The study group was staffed with about 450 people led on a day-to-day basis by Mark Craig, with an average of 250 people working directly on the study on any given day—although the core team was formed by the members of the AHWG.[30] The study began by decomposing the President's objectives into top level technology requirements. These requirements were then used to develop an end-to-end architecture, which included the following features:

- Characterize the environment in which humans and machines must function with robotic missions
- Launch personnel and equipment from Earth
- Exploit the unique capabilities of human presence aboard the Space Station Freedom
- Transport crew and cargo from Earth orbit to lunar and Mars orbits and surfaces
- Conduct scientific studies and investigate *in-situ* resource development

---

[27] O'Handley interview.

[28] Ibid.

[29] Ibid.

[30] Craig interview, 12 September 2003; Cohen interview, 9 December 2004.

The TSG assumed the agency would utilize the Space Shuttle and Space Station Freedom to implement SEI. This, in essence, meant the group never considered whether leveraging these systems was feasible or desirable given the existing fiscal environment. The inclusion of the two systems was almost a foregone conclusion because JSC wanted to protect the Shuttle and continue Station development—in the near term, this meant the ultimate success of SEI was not necessarily the agency's top priority. From the agency's perspective, completion of an orbital station was part of a serial progression that started with the shuttle and would eventually end with a human mission to Mars—an idea that dated back to post-Apollo planning. This viewpoint was directly influenced by Admiral Truly's decision to base the 90-Day Study's technical analysis on past NASA studies. Douglas O'Handley argues, "this is where the Space Council and the agency were on a collision course. NASA was documenting the past and the Space Council wanted options and innovative thinking. None of the NASA principals knew how to go about" providing those alternatives.[31]

The 90-Day Study alternative generation process was far from optimal. Because the TSG was so JSC-centric, technical and architectural concepts from other segments of the space policy community were not solicited. Perhaps more importantly, the group considered budgetary constraints last. This should have been the first thing that was evaluated, with all programmatic options tailored to the fiscal realities. Instead, the TSG put together a virtual 'wish list' for human exploration without taking into account the existing political environment. This eventually became an even greater problem because the group never paid "much attention to lowering the initiative's costs by using emergent technologies."[32] There is some indication that part of the reason for this was because NASA had been directed to virtually guarantee the safety of the astronauts. Based upon the Apollo experience and a contemporary understanding of the life science challenges, the TSG had calculated that one member of a seven-person crew may not return. The Space Council staff told agency planners they wanted 'seven out and seven back.'[33] This would have required 99.9999% mission reliability. As much as anything done by the space agency, this

---

[31] Report of the 90-Day Study on Human Exploration of the Moon and Mars, National Aeronautics and Space Administration, 20 November 1989, National Aeronautics and Space Administration Historical Archives, 2-2 to 2-5; NASA Administrator Richard Truly to Vice President Dan Quayle, 5 September 1989, Bush Presidential Records, George Bush Presidential Library; Craig interview, 12 September 2003; O'Handley interview.

[32] O'Handley interview.

[33] Ibid.

Chapter 4: The 90-Day Study

White House decision drove costs up enormously.[34]

SEI Takes Shape

In early November, the *Report of the 90-Day Study on Human Exploration of the Moon and Mars* began circulating at the Space Council.[35] The cover letter attached to the report stated the purpose of the study was intended as a data source for the Space Council to refer to as it considered strategic planning issues related to SEI. The document purported not to contain any official recommendations or estimates of total mission cost.[36] The preface made it abundantly clear that the TSG regarded President Bush's announcement speech as the initiative's guiding policy directive. As a result, the key doctrine that emerged from the report was expressed as follows. "The five reference approaches presented reflect the President's strategy: First, Space Station Freedom, and next, back to the Moon, and then a journey to Mars. The destination is, therefore, determined, and with that determination the general mission objectives and key program and supporting elements are defined. As a result, regardless of the implementation approach selected, heavy-lift launch vehicles, space-based transportation systems, surface vehicles, habitats, and support systems for living and working in an extraterrestrial environment are required." The analytic team did not include any alternative paths, but chose to strictly interpret Bush's announcement speech. This dogmatic approach was carried through the entire report, with a predictable outcome—a set of reference approaches requiring a massive in-orbit infrastructure and large capital investments.[37]

---

[34] On 2 November, President Bush signed National Space Policy Directive 1—a slight revision of the policy issued by the Reagan Administration 20 months earlier. The expansion of human presence and activity "beyond Earth orbit into the solar system" remained one of the nation's primary goals in space. Considering the administration's desire to have SEI provide a long-term direction for the American space program, however, the document didn't place a great deal of emphasis on the new initiative. Within the section dealing directly with civil space policy, human exploration was relegated to the bottom of a list of stated objectives for NASA—with Earth science, space science, technology development, and space applications at the top of the list. Even when addressing human exploration more specifically, the policy highlighted completion of Space Station Freedom and downplayed human missions beyond Earth orbit. Finally, the directive provided no specific guidance with regard to implementing the Moon-Mars initiative. [National Space Policy Directive 1, National Space Council, 2 November 1989, Bush Presidential Records, George Bush Presidential Library; Press Release, The White House Office of the Press Secretary, 16 November 1989, Bush Presidential Records, George Bush Presidential Library.]

[35] It was not officially released until 20 November 1989.

[36] Report of the 90-Day Study on Human Exploration of the Moon and Mars, National Aeronautics and Space Administration, 20 November 1989, National Aeronautics and Space Administration Historical Archives, cover letter.

[37] Ibid., Preface.

Mars Wars

To achieve the objectives set out in President Bush's announcement speech, the TSG adopted an evolutionary 30-year plan for SEI implementation. As the AHWG had done before it, the group put forth a strategic approach that depended on Space Station Freedom and followed initial human missions to the Moon and Mars with phased development of permanent human outposts on these celestial bodies—starting with emplacement, continuing with consolidation, and finishing with operations. Unlike the briefing that had been prepared during the agenda setting process, the 90-Day Study included a highly detailed description of NASA's vision for the robotic, lunar, and Martian phases of exploration beyond Earth orbit.[38]

Space Station Freedom
(Source: 90-Day Study)

The initiative would begin with precursor robotic missions intended to "obtain data to assist in the design and development of subsequent human exploration missions and systems, demonstrate technology and long communication time operation concepts, and dramatically advance scientific knowledge of the Moon and Mars." The TSG developed a logical progression of robotic explorers to address specific operational and scientific priorities. First, a Lunar Observer program would launch two identical flight systems on one-year polar mapping missions. Second, a Mars Global Network program would launch two identical flight systems carrying orbiters and multiple landers to provide high-resolution surface data at several locations. Third, a Mars Sample Return program would launch two identical flight systems to return five kilograms each of Martian rocks, soil, and atmosphere to Earth—this was the centerpiece of the robotic sequence. Fourth, a Mars Site Reconnaissance Orbiter program would launch two orbiters and two communications satellites to characterize landing sites, assess landing hazards, and provide data for subsequent rover navigation. Finally, up to five Mars Rover missions would certify three sites to determine the greatest potential for piloted vehicle landing and outpost establishment.[39]

---

[38] Ibid., section 3, The Human Exploration Initiative.

[39] Report of the 90-Day Study on Human Exploration of the Moon and Mars, section 3, The Human Exploration Initiative.

Lunar Transfer Vehicle would have provided transportation between *Space Station Freedom* and lunar orbit (Source: 90-Day Study)

As robotic exploration of the red planet was ongoing, the TSG strategy called for the development of a permanent lunar outpost. The mission concept for achieving this goal was highly complicated, relying on a vast in-orbit infrastructure, numerous spacecraft, and multiple resource transfers. The plan called for "two to three launches of the lunar payload, crew, transportation vehicles, and propellants from Earth to Space Station Freedom. At Freedom, the crew, payloads, and propellants are loaded onto the lunar transfer vehicle that will take them to low lunar orbit. The lunar transfer vehicle meets in lunar orbit with an excursion vehicle, which will either be parked in lunar orbit or will ascend from the lunar surface, and payload, crew, and propellants are transferred. [Then] the excursion vehicle descends to the lunar surface." A combination of cargo and piloted flights (with four crew members) would be utilized to construct the lunar outpost. The emplacement phase would begin with two cargo flights to deliver the initial habitation facilities, which included a habitation module (to be covered with lunar regolith to provide radiation shielding), airlock, power system, unpressurized manned/robotic rover, and associated support equipment. Emplacement would prepare the way for extended human missions during the consolidation phase, which would include erection of a constructible habitat to provide additional living space and experimentation with *in situ* resource utilization.[40]

The final step in the TSG strategy was the establishment of a human outpost on Mars. Similar to the lunar program, the Martian sequence would begin with the launch of the crew, surface payload, transportation vehicles, and propellant from Earth to Space Station Freedom. In LEO, the transfer and excursion vehicles would be inspected before setting out on the long journey toward the red planet. "Upon approach to Mars, the transfer and excursion vehicles separate and perform aerobraking maneuvers to enter the Martian atmosphere separately. The vehicles rendez-

---

[40] Report of the 90-Day Study on Human Exploration of the Moon and Mars, section 3, The Human Exploration Initiative.

Inflatable lunar habitat would have been outpost for up to 12 astronauts (Source: 90-Day Study)

vous in Mars orbit, and the crew of four transfers to the excursion vehicle, which descends to the surface using the same aero-brake. When their tour of duty is complete, the crew leaves the surface in the ascent module of the Mars excursion vehicle to rendezvous with the transfer vehicle in Mars orbit. The transfer vehicle leaves Mars orbit and returns the crew to Space Station Freedom."[41] Standard mission profiles for crewed flights to Mars would follow two different trajectory classes: one for a 500-day roundtrip with surface stays up to 100 days and one for a 1,000-day roundtrip with surface stays of approximately 600 days. After initial emplacement, the consolidation phase would entail assembly of a constructible habitat and utilization of a pressurized rover for long-range surface exploration.[42]

As envisioned by the TSG, implementation of SEI would require the construction of a new launch vehicle and multiple spacecraft to travel beyond Earth orbit. The study introduced two primary concepts for a heavy launcher, one a Shuttle-derived alternative and the other based on the proposed Advanced Launch System.[43] As indicated above, the in-space transportation system consisted of transfer and excursion vehicles—these systems would utilize chemical propulsion, although the report called for research funding to investigate nuclear propulsion. For Mars exploration, the transfer vehicle would actually carry the excursion vehicle to the red planet utilizing a large trans-Mars injection stage. The transfer vehicle would

---

[41] For cargo flights, an integrated configuration of two excursion vehicles is launched. Upon approach to Mars, the two vehicles separate and enter Mars orbit using aero-brakes. The first cargo flight in the Mars outpost mission sequence delivers the habitat facility to the outpost site.

[42] Report of the 90-Day Study on Human Exploration of the Moon and Mars, section 3, The Human Exploration Initiative.

[43] The Advanced Launch System (ALS) emerged in the mid-1980s as the rocket that would be used to deploy the space-based elements of the Strategic Defense Initiative program. However, by late 1989, it had become increasingly apparent that the requirements for the ALS program had largely disappeared. The initial phase of SDI would be deployed using existing Titan 4 and Atlas 2 rockets, and the launch requirements for subsequent phases of SDI deployment were too vague to require immediate development of ALS.

Mars Transfer Vehicle would have propelled crew and mars excursion vehicle to Mars orbit (Source: 90-Day Study)

Mars Excursion Vehicle would have transported four astronauts and 25-tons of cargo to Martian surface (Source: 90 Day Study)

include a crew module that would be a "single, pressurized structure 7.6 meters in diameter and 9 meters in length with…a life support system that recycles water and oxygen. The crew is provided private quarters, exercise equipment, and space suits that are appropriate for the long mission duration." The excursion vehicle crew module would provide living space during descent, ascent, and for up to 30 days in case of problems with the surface habitat.[44]

The TSG developed similar planetary surface systems for both Moon and Martian missions. In fact, the main rationale for development of a lunar outpost was as a testing ground for subsystem technologies for later missions to the red planet. The initial habitats for both outposts would be horizontal Space Station Freedom-derived cylinders 4.45 meters in diameter and 8.2 meters long. Laboratory modules would be attached to add expanded living volume. Each of these habitats would have regenerative life support systems capable of recovering 90% of the oxygen from carbon dioxide and potable water from hygiene and waste water. During the consolidation phase, an expanded habitat would be required to accommodate large crews and longer stays by providing more space. This would be a "constructible [11 meter] diameter inflatable structure partially buried in a crater or a prepared hole. This structure is an order of magnitude lighter than multi-module configurations of equivalent volume. Its internal structure includes self-deploying columns that

---

[44] Report of the 90-Day Study on Human Exploration of the Moon and Mars, section 3, The Human Exploration Initiative.

telescope upward and lock into place when the structure is inflated. When fully assembled and outfitted, the constructible habitat provides three levels, and has the volume required for expansion of habitat and science facilities. Major subsystems of the constructible habitat include the life support and thermal control systems, pressure vessels and internal structure, communications and information management systems, and interior outfitting." During this stage, a 100- kilowatt nuclear dynamic power system would begin providing the growing outpost much needed electric power (the plan called for ongoing progression of this capability, leading to a 550-kilowatt system). Initial surface exploration would be conducted using electric powered, unpressurized rovers. These vehicles would only have a range of 50 kilometers with human occupants, although they could be telerobotically operated for missions up to 1,000 kilometers from the outpost. This provided very limited capacity for long-range human exploration, which was nominally the primary reason for making the journey.[45]

The TSG mission plan was designed as the framework for selection of an overall "reference approach." The 90-Day Study included five different reference approaches, which were intended to provide different options (using only one mission strategy) for achieving President Bush's goals. The report introduced a set of metrics (cost, schedule, complexity, and program risk) that could be used by policy makers to decide the appropriate timeframe for SEI implementation. The reference approaches simply altered these metrics to provide different milestones for a single strategic plan. Thus, instead of examining numerous technical, operational, or strategic alternatives, the TSG chose to put forward one basic system architecture with slight timeline modifications. The different reference approaches included:

- **Reference Approach A**: Formulated to establish human presence on the Moon in 2001, using the lunar outpost as a learning center to develop the capabilities to move on to Mars. An initial expedition to Mars would allow a 30-day stay on the surface, with the first 600-day visit in 2018.
- **Reference Approach B**: A variation of Reference Approach A, which advanced the date of the first human Mars landing to 2011. This would reduce the ability to use the lunar outpost as a learning center for the Mars outpost.
- **Reference Approach C**: A variation of Reference Approach A, which advanced even further the date of the first Mars mission, but maintained the same expansion schedule for Mars outpost development.
- **Reference Approach D**: A variation of Reference Approach A, which

---

[45] Ibid.

Chapter 4: The 90-Day Study

slipped all major milestones two to three years.
- **Reference Approach E**: Formulated to reduce the scale of lunar outpost activity by using only a human-tended mode of operation and limiting the flight rate to the Moon to one mission per year. Three expeditionary missions to Mars (with 90-day surface stays) would precede the 2027 establishment of a permanent outpost with 600-day occupancy.

In essence, this set of reference approaches provided two limited alternatives (Reference Approaches A and E). The only difference between the two was the magnitude of lunar development and the timing of different milestones. There were no alternatives provided that suggested that it was not feasible from a budgetary perspective to attempt both a permanent return to the Moon and human exploration of Mars. In addition, as pointed out above, there were no alternatives that were based on significantly different mission profiles or technical systems. This represented a major shortcoming of the report, which would come back to haunt the space agency in subsequent months and years.[46]

The 90-Day Study included a cost estimate for the TSG's vision of SEI. It was based on a 30-year planning horizon and employed historical experience to "derive the approximate values for supporting development, systems engineering and integration, program management, recurring operations, new facilities, and civil service staffing levels." The TSG performed a parametric cost analysis using three regression models developed at different NASA field centers. The Marshall Space Flight Center Cost Model consisted of subsystem level data gathered from past human spaceflight programs, which was employed to estimate space transportation vehicle costs as a function of mass (assigning each reference approach a subjective complexity factor). The Johnson Space Center Advanced Mission Cost Model used a broader dataset drawing on developmental program statistics from NASA and other technology organizations to calculate expected surface system costs. Finally, the Jet Propulsion Laboratory Project Cost Model estimated program costs for robotic missions drawing on past analogous mission figures.[47]

The study provided funding estimates for reference approaches A and E, appraising expected costs from 1991 to 2025 (in constant fiscal year 1991 dollars). The estimates included reserves that accounted for nearly 55% of predicted expenditures, which was intended to allow for programmatic uncertainties. The report included tables that detailed the cost estimates for both reference approaches, separated into key phases:

---

[46] Ibid., section 4, Reference Approaches.

[47] Ibid., Cost Summary.

- **Reference Approach A**
  - Lunar Outpost: $100 billion (FY 1991-2001)
  - Lunar Outpost Emplacement & Operations: $208 billion (FY 2002-2025)
  - Mars Outpost: $158 billion (FY 1991-2016)
  - Mars Outpost Emplacement & Operations: $75 billion (FY 2017-2025)
  - Total: $541 billion
- **Reference Approach E**
  - Lunar Outpost: $98 billion (FY 1991-2004)
  - Lunar Outpost Emplacement & Operations: $137 billion (FY 2005-2025)
  - Mars Outpost: $160 billion (FY 1991-2016)
  - Mars Outpost Emplacement & Operations: $76 billion (FY 2017-2025)
  - Total: $471 billion

The report also included two startling charts, which illustrated the impact of the reference approaches on the overall NASA budget. Starting with a base budget of approximately $15 billion, the implementation of both reference approaches would require increasing the annual agency appropriation to $30 billion by FY 2000, where it would stay for another 25 years.[48] In the coming weeks and months, it would become increasingly clear that these budgetary requirements were simply staggering to all outside observers. Admiral Truly and the TSG clearly believed that President Bush was prepared to support a major escalation in annual spending for the space program. This judgment was reached despite the fact that the nation was facing large budget deficits and almost every other sector of the government was expecting significant funding cuts. It proved to be a tremendous miscalculation.

Mars Wars

Behind closed doors, the White House's reaction to the 90-Day Study was outright shock. The Space Council staff could not believe the TSG had produced a report that essentially had no real alternatives.[49] Mark Albrecht recalled later that the report included "basically one architecture…different technologies did not exist

---

[48] Ibid.

[49] Bryan Burrough, *Dragonfly*, p. 241.

at all, it was one plan offered three ways; slow, moderate, and fast. We were just stunned, felt completely betrayed. Vice President Quayle was furious. The 90-Day Study was the biggest 'F' flunk, you could ever get in government. The real problem with the NASA plan was not that we didn't think the technology was right, but that it was just the most expensive possible approach. It was just so fabulously unaffordable, it showed no imagination."[50] OMB Director Darman later told Congress, "some of us in the administration felt that the NASA report itself was very much biased towards what you might think of as the off-the-shelf approach to the Moon and Mars, that it didn't really seek highly divergent new technologies." When asked about the study after he left office, President Bush recalled feeling that "I got set up."[51]

The release led to a rapid disintegration in the already tenuous Space Council-NASA relationship. Douglas O'Handley remembers the study "made the situation worse than it was at the beginning. The NASA plan was not what the Space Council wanted. Admiral Truly lost all credibility with the Space Council. There was clearly a clash of personalities."[52] Although the White House was highly critical of NASA, agency officials believed the report was received unfavorably because of poor guidance from the Space Council. Admiral Truly and Aaron Cohen felt the council staff didn't really understand the technical complexities involved in going to the Moon and Mars. In their opinion, significantly different cost profiles were not needed because establishing a permanent human presence on those celestial bodies would require approximately the same amount of resources, regardless of the strategic architectures that were selected. If the Space Council wanted options with different budgetary impacts, they argued, NASA should have been asked to examine different mission content—such as eliminating construction of a permanent lunar base or a human mission to Mars.

In retrospect, this argument was somewhat disingenuous considering it was Admiral Truly who had originally argued that any new initiative should include a permanent return to the Moon <u>and</u> human missions to Mars. The White House had initially wanted to consider simply returning to the Moon, although Quayle and Albrecht went along with the space agency's revised plan and adopted it for President Bush's announcement speech. In any case, it was NASA's feeling that short of changing mission content, it would be nearly impossible to reduce significantly the budgetary requirements. This conclusion was reached at least partially because

---

[50] Albrecht interview.

[51] Warren E. Leary, "Plans for Space Are Realistic, Official Says," *New York Times* (17 December 2003).

[52] O'Handley interview.

the administration was believed to be highly risk averse, which increased programmatic costs. Additionally, the TSG didn't believe that approaching technological changes more aggressively would reduce costs markedly.[53]

NASA leaders were actually surprised with the White House reaction to the 90-Day Study. Even after submitting it, they believed that it was an appropriate response to President Bush's speech. This was particularly true regarding the cost estimates for a long-term initiative. The TSG believed that given Quayle's approval of similar budget estimates in June, these new figures were acceptable. Years later, Aaron Cohen concluded that the primary reason for this misunderstanding was poor communication between the two organizations. He maintained that the primary problem was not that the council staff asked for options, but that "NASA did not try to understand what the customer really wanted. That is a very important rule in design. Know what the customer really wants. The president's speech was not what they really wanted. We should have tried harder to understand what they really wanted."[54] From his point of view, the primary lesson learned from this failure of communication was that technical agencies like NASA need to work very closely with its customer to understand their policy needs.[55]

Before the public release of the report, Vice President Quayle and Mark Albrecht began preparing a plan to downplay the importance of the 90-Day Study. The Space Council staff fashioned a strategy intended to discourage speculation that the TSG vision for SEI matched that of President Bush. The Administration would argue the study was only one data source within an ongoing alternative generation process that would seek inputs from other government agencies, industry, and the scientific community. In fact, such a process had not yet been initiated. Regardless, the White House would suggest that the TSG report was merely a starting point for identifying ways to minimize risk, maximize performance, keep costs at an affordable level, and achieve overall program goals.[56] This was clearly an ad hoc effort at damage control. It proved to be a near total failure.

The clearest sign that the Space Council staff was on high alert was the decision to ask NASA to remove the cost estimation section from the public report. Mark Albrecht recalled this determination was made because "we had only one alternative, and it had this $400 billion price tag. We knew right away that this was

---

[53] Cohen interview.

[54] Ibid.

[55] Ibid.

[56] Talking Points, NASA Moon/Mars Database Report, National Space Council, 14 November 1989, Bush Presidential Records, George Bush Presidential Library; Albrecht interview.

dead on arrival. Just what you need, to go out and say President Bush's initiative is going to cost $400 billion. Dead. That's exactly what we didn't want to happen. So, we asked NASA to keep it separate while we began in earnest to look for real alternatives." Regardless, the administration correctly concluded that rumors would quickly surface that the agency had indeed conducted an internal cost analysis. To counter these stories, the council staff planned to cast doubt on the TSG's estimation techniques—most importantly challenging the group's conservative technical approach.[57]

In mid-November, the White House scheduled a meeting of the full Space Council to provide NASA with an opportunity to officially brief the 90-Day Study. This meeting would prove to be one of the seminal events in the history of SEI. In the days leading up to the meeting, the Space Council staff began searching for alternative approaches for Moon-Mars exploration. In particular, they were interested in an architecture developed at Lawrence Livermore National Laboratory (LLNL) by a team led by physicist Lowell Wood. The concept was designed to put humans on Mars in ten years, for $10 billion, using inflatable modules to build spacecraft and Martian bases. Although this option was considered technically risky (and would probably cost more than Livermore estimated), Vice President Quayle was eager to investigate inventive approaches that differed from those proposed by the TSG. Two days before the council meeting, Mark Albrecht called Wood and asked "how fast can you get your concept printed up into a document." The reply was overnight. On 16 November, Albrecht contacted Kathy Sawyer from *The Washington Post* and told her that two different mission architectures would be briefed at the council meeting the following morning—one from NASA and one from LLNL. He made it clear that the administration would be searching for additional technical alternatives for accomplishing the Moon-Mars mission.[58]

The next day, *The Washington Post* ran a story indicating that Vice President Quayle had decided to end NASA's monopoly on developing concepts for space exploration. The piece indicated that Quayle was greatly concerned with the agency's big budget approach to SEI and wanted Congress to know that the administration would be considering other options. White House aides were quoted saying he had directed the Space Council to conduct an "open competition for money-saving ideas for human exploration of the Moon and Mars."[59] While this was clearly designed

---

[57] Ibid.

[58] Albrecht interview.

[59] Kathy Sawyer, "Quayle to Give NASA Competition on Ideas for Space Exploration," *The Washington Post* (17 November 1989).

to assuage Congressional concerns, it was also an admission that the Council had initially abdicated this responsibility. The article highlighted the LLNL concept as one potential alternative. Douglas O'Handley, who up until that point had generally been sympathetic toward the Space Council, was quoted as saying, "I'm amazed it would be taken…seriously. If it is, it's by an individual that doesn't understand space risks."[60] O'Handley believed the LLNL concept was not even worth discussing because it displayed a fundamental lack of knowledge. Within the space policy community, consideration of the Livermore architecture raised serious questions regarding the technical credibility of the Space Council staff.[61]

That morning the Space Council met at the White House. When the members entered the room, the LLNL study was on everybody's seat. A senior White House aide who attended the meeting recalled, "the NASA guys were absolutely furious, and we were furious because they'd tried to screw us, and we screwed them right back."[62] After opening the meeting, Vice President Quayle asked NASA and LLNL to brief their respective studies. After the presentations were made, Dick Darman, James Watson (Secretary of Energy), and Donald Rice (Secretary of the Air Force) asked Admiral Truly a series of pointed technology-related questions. The thrust of these queries centered on whether SEI was politically viable or possible given the budgetary realities without breakthrough developments in technology. The Council's position was clearly that NASA needed to aggressively force some technologies through the system.[63] As a result, the group decided to ask the National Research Council to conduct a review of the 90-Day Study and search for possible alternatives.[64] With the economy in a rut and a growing budget deficit, the Council felt the Administration needed to investigate cheaper options. Thus, the most significant outcome of the meeting was the displacement of NASA as the agency responsible for developing technical approaches for the initiative. Mark Albrecht recalled, "At

---

[60] Ibid.

[61] O'Handley interview.

[62] Senior Administration Official interview via electronic-mail, 5 November 2003.

[63] James Fisher and Andrew Lawler, "NASA, Space Council Split Over Moon-Mars Report," *Space News* (11 December 1989), p. 10.

[64] On 4 December, Vice President Quayle sent a letter to Dr. Frank Press, Chairman of the National Research Council (NRC), officially requesting that his organization conduct a review of the 90-Day Study. Quayle requested that the NRC consider alternative approaches, or a range of options, for human exploration of the solar system. He included a list of questions that he hoped the NRC would address in its review, focusing on whether the 90-Day Study addressed the widest range of technically credible approaches for implementing SEI. The letter concluded by requesting that the NRC complete the review by the end of February 1990. [Vice President Quayle to Dr. Frank Press, 4 December 1989, Bush Presidential Records, George Bush Presidential Library.]

Chapter 4: The 90-Day Study

that meeting, the Space Council basically took over the SEI project."[65] In retrospect, however, one must ask why this authority was ever in question.

When the 90-Day Study was released publicly, the Congressional reaction closely matched that of the Space Council—which did not bode well for SEI. For Dick Malow, one of the key problems with the TSG approach was that it only called for a 30-day surface stay for initial Martian landings. Based on this short time period for scientific exploration, he remembered thinking the costs seemed horrifically high. "I think the idea of only spending 30 days on the Martian surface drove me gradually to the feeling that SEI was a mistake and that it wouldn't sell. I couldn't see how it made sense to spend $400 billion for a 30-day stay on Mars."[66] There was a sense on Capitol Hill that the study was basically Johnson Space Center's position on how the initiative should be carried out and that it was ultimately doomed. Malow recalled thinking even before the study was released that there was "just something totally out of whack. When you coupled the cost of the initiative with the problems ongoing with the space station, it just seemed like the wrong time and the wrong approach. The study only reinforced that view."[67] Fellow staffer Stephen Kohashi shared this opinion and felt that the Administration never truly recovered momentum after the report was released. "The biggest problem was the cost of the Space Station development, as well as the cost of maintaining development schedules for other on-going projects. With on-going projects getting cut back because of overall budgetary constraints, initiation of a manned Mars mission, with its enormous run-out costs, seemed fanciful at best."[68] As a result, the report lacked all credibility as a serious proposal in Congress.

The timing of the study's release, combined with its poor reception, could not have been worse for the space agency. During this period, the White House was working on its fiscal year 1991 budget. NASA had proposed new funding to develop the necessary technology base to undertake future missions to the Moon and Mars. The initial budget submission included an SEI-specific increase of $450 million. There were also SEI-related increases to on-going projects. In late November, in the aftermath of NASA's unveiling of the 90-Day Study, the OMB passback cut

---

[65] Albrecht interview.

[66] Malow interview.

[67] Ibid.

[68] Kohashi interview.

nearly $270 million from the SEI-specific request.[69] The Administration realized that given the Congressional reaction to the report, it had to significantly reign in the agency's efforts to obtain a large budget increase. The hope was that this would give the administration time to reorient the initiative and build momentum for its implementation among key members of Congress. The $188 million in SEI-specific funds that were ultimately included in the President's budget were intended to commence technology efforts deemed crucial for eventual Moon-Mars missions, including:

- Space Transportation Capability Development
  – Space Transportation Main Engine: develop a low-cost advanced main engine propulsion system critical to any advanced transportation system ($40M)
  – Heavy-lift Technologies: define advanced transportation technology development activities for a heavy-lift launch vehicle to support lunar and Mars missions ($10M)
- Space Science
  – Mars Observer: extend planned mission duration and upgrade image processing capability to provide more detailed topographical data to determine potential landing sites for future Mars missions ($15M)
  – Lunar Observer: new mission to provide detailed data on the moon's topography, geology, and climatology to assist in determining potential landing sites of optimal scientific merit ($15M)
- Space Research & Technology
  – Exploration Technology Increment: develop applications such as atmospheric aero-braking, advanced automation and robotics, space nuclear power, space transfer vehicle propulsion, cryogenic fuel transfer and conservation, regenerative life support systems, radiation protection shielding, high-rate optical frequency and Ka-band communications, and science multi-spectral sensors ($88M)
- Space Station
  – Advanced Programs Increment: assess and do preliminary designs of high-pressure space suit, solar dynamics power, and advanced propulsion system for reaction control ($20M)[70]

---

[69] Admiral Richard Truly to Richard G. Darman, 27 November 1989, Library of the National Aeronautics and Space Administration Chief Financial Officer.

[70] National Aeronautics and Space Administration, "Budget Estimates: Fiscal Year 1991, Volume 1," Library of the National Aeronautics and Space Administration Chief Financial Officer.

Chapter 4: The 90-Day Study

Combined with approximately $140 million in new funds for ongoing programs, these budget augmentations were intended to be the first steps in the development of the necessary technology to undertake human exploration beyond Earth orbit.

At the end of November, the Space Council invited a Blue Ribbon Discussion Group to the White House to provide it with advice regarding the administration's approach to SEI. Mark Albrecht recalled the council staff was trying to figure out how to get what "we asked for, given that NASA has completely flunked the course."[71] The panel included such space policy luminaries as astronomer Carl Sagan, physicist Edward Teller, former astronaut and U.S. Senator Harrison Schmitt, former astronaut Michael Collins, former Air Force Secretary Pete Aldridge, and former NASA Administrator Thomas Paine. The group spent its first day and a half receiving briefings on a wide variety of topics, ranging from technology to international cooperation to program rationales. NASA presented the 90-Day Study. Lowell Wood briefed the LLNL plan. Finally, Boeing shared its plans for development of heavy-lift expendable launch vehicles and next-generation human spacecraft.[72]

During the final afternoon of the two-day summit, Dr. Laurel Wilkening of the University of Washington provided Vice President Quayle, Budget Director Richard Darman, and Science Advisor Allan Bromley with an overview of the group's findings and recommendations. First, the administration had not outlined an adequate rationale for SEI. Without a more compelling justification for the initiative, the administration would be unlikely to gain long-term public and congressional support for the undertaking. Second, NASA should be directed to prepare a broad range of technical options for presidential consideration, drawing on external and internal sources outside JSC. The 90-Day Study reference approaches were seen as "business as usual" approaches that did not seek to leverage breakthrough technologies. Third, NASA should embark on the rapid development of advanced heavy launch vehicles as the most important initial implementation step. Fourth, Tom Paine and others argued NASA's managerial structure was not well suited to carry out the program. They suggested that a new 'exploration agency' should be created within NASA, likely requiring a new field center separate from the existing operational side of the organization. Finally, the administration should evaluate opportunities for inter-

---

[71] Albrecht interview.

[72] Schedule Proposal, Mark Albrecht to CeCe Kramer, 9 November 1989, Bush Presidential Records, George Bush Presidential Library; Memorandum, Mark Albrecht, 30 November 1989, Bush Presidential Records, George Bush Presidential Library; Andrew Lawler, "Panel: Rationale Missing for Moon-Mars Proposal," *Space News* (11 December 1989); Brad Mitchell to Andy Card, 4 December 1989, Bush Presidential Records, George Bush Presidential Library.

national cooperation that would not make the space program dependent on other nations for mission essential activities.[73]

By mid-December, the clash underway within the Bush administration regarding the future direction of the American space program had become public knowledge. *The Washington Post* described this out of sight "Mars Wars," as an on-going conflict between the Space Council and NASA, characterized by communications failures, disagreements, and outright turf battles. As had been the case for many months, the Council's continued appeal for a more imaginative technical approach for implementing SEI caused friction. The White House position, supported most importantly by Vice President Quayle and OBM Director Darman, was that SEI should be used as an opportunity to seek economic benefits from directed technology investments. This would have the added benefit, they believed, of dramatically reducing the costs of future Moon-Mars missions. Quayle and Darman were clearly disappointed with the 90-Day Study's failure to cast a wider net in search of new technologies. They believed the TSG plan was simply bureaucratic business as usual. On 1 December, Admiral Truly and Mark Albrecht had had a private confrontation at NASA headquarters when the latter expressed this White House view. Although the meeting was downplayed publicly, there was clearly a heated conversation regarding the appropriate course of action for SEI.[74] An article in *Space News* captured the essential dilemma SEI posed for both the administration and NASA. "Administration officials have privately chastised NASA for preparing a report that some felt was unimaginative and did not adequately address new technologies that could dramatically cut costs for the venture. NASA officials have blamed the Space Council for the problems, saying that direction given to the agency was unclear."[75]

Regardless of the ongoing hostilities between the Space Council and NASA, the latter was working to salvage SEI. On 14 December, the full council met to discuss a staff report drafted after the conclusion of the two-day Blue Ribbon Discussion Group meeting. The report suggested that NASA should be instructed to issue a call to industry, universities, and research centers for high leverage system concepts based on new technology developments—including nuclear propulsion and innovative uses of existing technologies. Following an agency review of all inputs to provide a "sanity check," these alternatives would be forwarded to the NRC for more

---

[73] Ibid.

[74] Kathy Sawyer, "En Route to Space Goal, Groups Diverge: Friction Between NASA and Quayle's National Council Erupts in "Mars Wars,'" *The Washington Post* (11 December 1989).

[75] James Fisher and Andrew Lawler, "NASA, Space Council Split Over Moon-Mars Report," *Space News* (11 December 1989).

Chapter 4: The 90-Day Study

detailed analysis. This process was expected to take a year.[76] The panel also addressed the Blue Ribbon Discussion Group's concern regarding the need for a compelling rationale for SEI. The council members decided the space program should be portrayed as an investment that was greater than the sum of its parts, which provided the nation with a unique combination of benefits. The group selected the following framework to describe the administration's vision:

The Council believed that NASA should build on its proven track record "for calling forth factors which are critical to America's greatness—pushing the frontier, creating economic opportunity from adversity, and looking over the horizon to create a better world for tomorrow's generation."[77] In retrospect, this rationale was more of a propaganda statement than an actual description of the Bush administration's reasoning for supporting human exploration of the Moon and Mars. The real motivation for SEI was quite simple—provide direction to a directionless agency. While this justification may have been adequate to force a fundamental redirection of existing NASA programs, it was not compelling enough to gain Congressional support for an initiative that would require doubling the space agency's budget. In fact, this had never been the administration's intent. The Space Council's inability to control the alternative generation process, however, led to the release of a NASA report that necessitated a massive increase in funding.

Leadership / Pride	Economic Impact	Legacy
Explore new frontiers	Enjoy unanticipated benefits	Create a better world for next generation
Challenge best and brightest	Create new jobs and foster aerospace industry	
Foster standard of excellence in science education	Enhance economic competitiveness	
Manifest destiny	Promote science and medical innovation	

---

[76] Simon P. Worden to the National Space Council, "Strategic Planning for the Space Exploration Initiative: The How, What, and When?" 14 December 1989, Bush Presidential Records, George Bush Presidential Library.

[77] Courtney Stadd to Brad Mitchell, Ed McNally, and Joe Heizer, "Space Exploration Initiative," 18 December 1989, Bush Presidential Records, George Bush Presidential Library.

On 19 December, based on the recommendation of the Blue Ribbon Discussion Group, Vice President Quayle directed NASA to take the lead in seeking out and evaluating technical and policy alternatives for implementing SEI. Quayle wrote Admiral Truly saying "America's space program has always been a recognized source of innovation. We need to bring that same innovativeness to bear today [and] ensure that all reasonable conceptual space exploration alternatives have been evaluated. We need to cast our net widely drawing upon America's creative potential to ensure that we are benefiting from a broad range of promising new technologies, and innovative uses of existing technologies." Quayle's position was clearly that not only had the 90-Day Study been based on outdated technology, but the TSG had not even applied those technologies in innovative ways. This was harsh criticism. The decision to give the space agency competition for strategic planning for human exploration was unprecedented in the history of the space age. It removed a 30-year NASA monopoly and signaled a serious lack of confidence in the organization's ability to formulate pioneering mission designs while taking into account the challenges of the existing political environment.[78]

Two days later, a seemingly humbled Truly announced that "the time has come now, not only to continue our analysis of exploration mission alternatives but also to begin actual pursuit of the innovative and enabling technologies that have been identified as necessary to proceed." This statement was a very public admission that the TSG had not identified innovative approaches for SEI. Nevertheless, Truly stated his belief that the Office of Exploration had conducted exceptional work setting a foundation for future human missions to the Moon and Mars. This was perhaps an indication that NASA still felt that its vision represented the best option for exploration of the solar system.[79/80]

---

[78] Vice President Quayle to Admiral Richard Truly, 19 December 1989, Bush Presidential Records, George Bush Presidential Library.

[79] Press Release 89-185, National Aeronautics and Space Administration, 21 December 1989, Bush Presidential Records, George Bush Presidential Library; William J. Broad, "NASA Losing 30-Year Monopoly In Planning for Moon and Mars," *The New York Times* (15 January 1990).

[80] Admiral Truly did not formally reply to the White House direction until 31 January 1990. In a letter to Vice President Quayle, he provided details of a process for soliciting outside strategic approaches for SEI implementation. This process would include the release of a NASA Research Announcement. The space agency would specifically seek inputs from professional societies (including the American Institute of Aeronautics and Astronautics—AIAA) and other federal agencies. The plan also envisioned a national conference that would be jointly sponsored by NASA and AIAA. All of these efforts would be coordinated through a newly created Office of Aeronautics, Exploration, and Technology. [Admiral Richard Truly to Vice President Quayle, 31 January 1990, Bush Presidential Records, George Bush Presidential Library.]

Chapter 4: The 90-Day Study

Over the course of the subsequent month, the Space Council began working behind the scenes to get SEI back on track. This included Council meetings aimed at developing a policy directive for the initiative and conferences with key members of Congress. Most importantly, however, the Administration was working on its fiscal year 1991 budget. President Bush was set to request a dramatically increased space budget, 24% higher than the previous year. This included a significant commitment to Mission to Planet Earth (raised 53%), a considerable boost in funding for Space Station Freedom (raised 39%), and a noteworthy enlargement of the space exploration budget (raised 32%). The budget included a new account for the Moon-Mars initiative, but the administration intended to make it clear that for the next few years SEI-related efforts would focus on pushing emerging and new technologies—those with the greatest potential for achieving the initiative faster, better, and cheaper.[81] The White House did not intend to lock onto any single program design or architecture until a thorough search for innovative technical alternatives had been completed.[82]

In late January, when the budget was officially submitted to Congress, the new funding for SEI was grouped together with on-going exploration efforts. Combining these budget lines made it appear that the space agency was asking for approximately $1.3 billion to support the robotic science missions and research and development activities required to push the initiative forward. In reality, only $300 million represented new spending for programs associated with SEI.[83] After the budget's public release, *The New York Times* reported that the proposed 24% increase in the NASA budget (a $2.8 billion boost to $15.1 billion) signaled President Bush's commitment to put money behind the exploration initiative. Admiral Truly stated "this is the most important budget for us … it's going to set our course, if we're successful, for many years to come and over a wide range of programs."[84] The article pointed out

---

[81] A few years later, NASA Administrator Dan Goldin, who would become the second administrator appointed by the Bush administration, made "faster, better, cheaper" the mantra of NASA. The concept emerged earlier, however, as the administration was trying to "infuse that kind of SDI mentality" into the SEI alternative generation process. [Albrecht interview]

[82] Talking Points, Meeting with Republican Members of House Science Committee, National Space Council, 22 January 1990, Bush Presidential Records, George Bush Presidential Library.

[83] White House Office of Management and Budget, "Budget of the United States of American, Fiscal Year 1991," 29 January 1990, pp. 49-82.

[84] "NASA Budget Press Conference: Statement of Richard H. Truly, NASA Administrator," *NASA News* (29 January 1990); John Noble Wilford, "Budget for the Space Agency Sets Broader Course in Exploration," *The New York Times* (2 February 1990), p. 19.

that although the first year funding for SEI was modest, it would presumably multiply several times over coming years. John Pike from the Federation of American Scientists was quoted saying, "There's a danger with all these big projects just getting started. They could just about devour everything else in the space program."[85] This was clearly a congressional concern and easing this worry would be the focus of the administration in the coming months.

A month later, the National Research Council briefed the findings of its review of the 90-Day Study to the full Space Council. The fundamental conclusion of the 14-member review panel was that the TSG report had critical flaws and that additional options needed to be studied to provide decision-makers with a 'menu of opportunities.' Former Presidential Science Advisor H. Guyford Stever, who chaired the NRC committee, argued SEI warranted "more intense scrutiny and evaluation of options must take place before decisions are made regarding mission architecture." This was an apparent criticism of the rushed nature of NASA's internal analysis and the resultant lack of real options for policy makers to choose between. The committee also subtly reproved past White House actions, stating that three crucial questions needed to be answered before moving on to actual concept development. These questions included: what is the appropriate pace for SEI; what is the appropriate scope of SEI; and what level of long-term support will SEI receive.[86] These were clearly high-level policy questions under the Space Council's purview that arguably should have been answered before the space agency was allowed to embark on the 90-Day Study.

After examining the reference approaches presented by the TSG, the NRC panel concluded that the JSC-led team was overly risk averse in the area of technology development. While there is some evidence that this was the result of guidance from the White House, the committee believed advanced technologies offered opportunities for more rapid and cost-effective access to and exploration of space. The group suggested a balanced technical approach would provide risk reduction and management opportunities. To this end, it recommended that early technical work should be focused on developments in four areas:

- A new generation of heavy-lift launch vehicles to transport large cargoes to LEO
- A new, robust, and efficient reusable launch vehicle for transport of humans and precious cargo to LEO

---

[85] Ibid.

[86] Committee on Human Exploration of Space, *Human Exploration of Space: A Review of NASA's 90-Day Study and Alternatives* (Washington, DC: National Academy Press, 1990); Press Release, National Research Council, 1 March 1990, Bush Presidential Records, George Bush Presidential Library.

- Nuclear propulsion technologies for more rapid transport of human crews across the solar system
- Nuclear electric power to meet the high energy requirements of lunar and Martian outposts

The panel openly criticized NASA's failure to discuss the eventual phase-out of the Shuttle, which it deemed to be a crucial step in providing highly reliable, less labor-intensive launch operations. It further expressed its belief that NASA should have emphasized the importance of safe nuclear power and propulsion for SEI. The NRC found that these technologies were essential to meeting the mission's substantial electricity demands and to reduce weight and travel time to Mars. In conclusion, the committee made an appeal to Congress to reserve judgment regarding SEI until investigations had been carried out to examine the merits of different technical approaches that may prove more feasible in the existing political environment.[87]

The White House reaction to the NRC review was extremely positive. Mark Albrecht felt "it was independent verification that we're not dreaming in the search for viable and credible alternative architectures." There was relief among the council staff that the NRC had concurred with their conclusion that there was a limited amount of technical imagination in the 90-Day Study, which made the TSG approach very expensive. They were also pleased that the report indicated there were additional alternatives that could dramatically decrease the overall cost of implementing SEI. Mark Albrecht recalled the report "was just a huge boost in our credibility around town and inside the White House."[88] Even many at NASA felt the report was a fair criticism of the 90-Day Study, but even at this point, the space agency had no plans to expand the envelope at all.[89]

The release of the NRC review was a critical turning point in lifecycle of SEI. It represented an authoritative declaration that NASA's 90-Day Study had serious flaws and failed to provide policy makers with an adequate tool to decide the future course of the American space program. While it clearly criticized the space agency, it also implied that there had been a White House failure to provide necessary guidance with regard to the pace, scope, and long-term commitment to the initiative. This public revelation forced the administration to step up an ongoing process aimed at saving the program. Over the coming months and years, intensified efforts were made to make President Bush's vision of a permanent return to the Moon and human exploration of Mars a reality.

---

[87] Ibid.

[88] Albrecht interview.

[89] O'Handley interview.

5

The Battle to Save SEI

"And so as this century closes, it is in America's hands to determine the kind of people, the kind of planet, we will become in the next. We will leave the Solar System and travel to the stars. Not only because it is democracy's dream, but because it is democracy's destiny."

President George Bush, 11 May 1990

Throughout the fall of 1989, President Bush had not been heavily engaged in the evolution of SEI within his administration. He had largely delegated responsibility for the initiative to Vice President Quayle, while he addressed more pressing events on the international stage—most importantly, the virtual implosion of communism in Eastern Europe. International tensions remained a fact of life during the coming months as reunification efforts began in East and West Germany; independence movements gained momentum in several Soviet republics; President Gorbachev proposed that the Communist Party give up its monopoly on power in the U.S.S.R.; and Panamanian dictator General Manuel Noriega overturned democratic elections that had effectively ousted him from power. Regardless, during the early part of the new year, President Bush was able to return his attention to domestic matters—including the fate of the American space program.[1]

---

[1] John Robert Greene, *The Presidency of George Bush* (Lawrence, KS: University of Kansas Press, 2000), pp. 89-106.

Presidential Decisions

During the early months of the new year, the Space Council staff began developing actual policy directives for the implementation of SEI. Based on direction provided by the full Council during two meetings on the subject, the staff was tasked with drafting two documents. The first would provide general policy guidance, while the second would introduce a course of action for including international partners in Moon-Mars missions. In a sign that the Administration had lost complete faith in NASA, the staff turned to the Department of Defense to conduct most of the analytical work necessary to develop these documents. Although NASA leaders had originally been in favor of re-establishing the Space Council, this view had dramatically shifted now that the new organization had turned to the military to comment on and critique the space agency's plans and projects.[2] Regardless, over the period of several months, the council staff worked closely with military analysts and the 'deputies committee,' a group consisting of high-level representatives from each of the Council's member agencies, to gain a consensus on the wording of the forthcoming policy statements.[3]

On 21 February, President Bush signed a Presidential Decision on the Space Exploration Initiative. Fully supported by Vice President Quayle and the National Space Council, its public unveiling three weeks later was clearly timed to coincide with the release of the NRC review of the 90-Day Study. The NRC panels' findings and recommendations largely validated the policy guidance found within the presidential directive. The objective of the document was to provide the American space program with near-term guidance for carrying out the long-term SEI vision. The policy consisted of the following components:

- The initiative will include both lunar and Mars program elements
- The initiative will include robotic science missions
- Early research will focus on a search for new and innovative approaches and technologies
  - Research will focus on high leverage technologies with the potential to greatly reduce costs
  - Mission, concept, and systems analyses will be carried out in parallel with technology research
  - Research will lead to definition of two or more significantly different exploration architectures

---

[2] Albrecht interview.

[3] Mark Albrecht to National Space Council, 16 January 1990, Bush Presidential Records, George Bush Presidential Library; Mark Albrecht to National Space Council, 2 February 1990, Bush Presidential Records, George Bush Presidential Library.

- A baseline program architecture will be selected from these alternatives
• Three agencies will carry out the initiative, with the National Space Council coordinating all activities
  - NASA will be the principal implementing agency
  - DOD and DOE will have major roles in technology development and concept definition

Coming eight months after President Bush first announced the initiative, this directive provided the direction that had clearly been needed within a fractious policy making community.[4] It was among the most significant documents in the chronicle of SEI. It represented an outright skirmish in the battle to gain control of strategic space policy planning within the Bush administration. Mark Albrecht said later, "it took us almost a year to go where we wanted to go directly and it cost us time, it cost confusion on the Hill."[5] Although this was a criticism of NASA, the Administration itself shared equally in the blame for not providing the required direction earlier. It is unclear whether the ultimate fate of SEI would have changed even if policy guidance had been provided much earlier, but it seems safe to conclude that the lack of administration leadership significantly reduced the initiative's chances of success. By the time the council finally supplied the needed direction, it was probably too late to resurrect an undertaking Congress presumed would be outrageously expensive. Dick Malow recalled that even with of a presidential directive providing policy guidance for SEI, "the general feeling about the program on the Hill continued to weaken."[6]

At the end of March, the White House made public a second presidential directive announcing President Bush's decision to commence discussions with foreign nations regarding international cooperation for SEI. This idea had been encouraged the previous summer by Carl Sagan, who sought to take advantage of warmer relations between the United States and Soviet Union. The Department of Transportation's (DOT) Commercial Space Transportation Advisory Committee had likewise recommended cooperation with the U.S.S.R. The committee's chairman,

---

[4] Presidential Decision on the Space Exploration Initiative, National Space Council, 21 February 1990, Bush Presidential Records, George Bush Presidential Library; Press Release, The White House Office of the Press Secretary, 8 March 1990, Bush Presidential Records, George Bush Presidential Library; "Cold Water on Mars," *The Economist* (10 March 1990), pp. 94-95; "Bush Calls for Two Proposals for Missions to Moon, Mars," *Aviation Week and Space Technology* (12 March 1990), pp. 18-19; Memorandum, Mark Albrecht to Ed Rogers, 13 March 1990, Bush Presidential Records, George Bush Presidential Library.

[5] Albrecht interview.

[6] Malow interview.

Alan Lovelace, said "cooperation with the Soviets is logical given the great desire of the administration to take steps to support developments in Eastern Europe." Another potential reason to cooperate was to reduce the U.S. contribution to the expensive initiative. It was suggested that the issue should be placed on the table for the Bush-Gorbachev summit planned for the summer.[7] The policy document itself indicated that the nation should pursue negotiations with Europe, Canada, Japan, and the Soviet Union. It was believed that this decision directive would support three important objectives. First, and most importantly, it would expand the coalition of initiative supporters by adding a foreign policy rationale. Second, it would involve partners capable of contributing financial resources to an expensive undertaking. Third, it would involve partners with important technical capabilities—most notably Soviet experience addressing the impacts of prolonged spaceflight and constructing nuclear space systems.[8]

The Soviet reaction to President Bush's call for international cooperation was extremely positive. Four years earlier, President Gorbachev had asked President Reagan to join his nation in a joint mission to the red planet, which would have met a long held ambition within the U.S.S.R. for human space exploration focusing on a voyage to Mars. After the release of the presidential directive, the spokesman for the Soviet embassy in Washington stated, "we have always been for cooperation with the United States in this area." Despite this encouraging response, by the end of the month, the NRC panel that had been evaluating SEI publicly warned against any cooperative robotic sample return missions to Mars with the Soviets. While the NRC did not address human exploration, it found that a highly interdependent undertaking could make planetary science "a potential hostage to political events." In the long-run, the potential benefits sought from exploring international cooperation were never realized.[9]

While the Space Council was working to provide long overdue policy guidance for SEI's implementation, senior NASA officials were appearing on Capitol Hill to defend the proposed increase in the agency's budget. In late March, the House Appropriations subcommittee with authority over the NASA budget held two days of hearing on the matter. It became apparent very quickly that the committee, chaired by Representative Bob Traxler (D-MI), was committed to identify-

---

[7] "Bush Seen Cooperating with Soviets on Moon-Mars Project," *Dow Jones News Service* (18 January 1990).

[8] "International Cooperation in the President's Space Exploration Initiative," The White House, Office of the Press Secretary, 30 March 1990, Bush Presidential Records, George Bush Presidential Library.

[9] William J. Broad, "Bush Open to Space Voyages with Soviet Union," *The New York Times* (3 April 1990), sec.C2; Craig Covault, "White House Approves Soviet Talks on Moon/Mars Exploration Initiative," *Aviation Week & Space Technology* (9 April 1990), p. 24; James R. Asker, "NRC Warns U.S. Against Joint Missions to Mars With Soviets," *Aviation Week & Space Technology* (23 April 1990).

ing and eliminating all funding associated with SEI. Chairman Traxler began the SEI-related questioning by asking NASA if the net increase in spending associated with the initiative was approximately $300 million—if on-going programs such as the National Aerospace Plane, Space Station Freedom, and Mars Observer were not included. NASA Comptroller Thomas Campbell answered that this was correct. Traxler then asked a series of questions regarding the technologies included in the 90-Day Study, which led to a long response from Admiral Truly defending the report—he concluded that the $188 million in funding for new technologies were dedicated to ascertaining what innovations would be required to efficiently explore the Moon and Mars. An undeterred Traxler followed-up by asking whether Truly agreed with the NRC report, which concluded that SEI as envisioned in the 90-Day Study would be technically challenging and very expensive. After Truly answered in the affirmative, Traxler got to the heart of the Congressional concern by asking whether fully implementing the TSG plan would require more than doubling the NASA budget. Once again answering affirmatively, Truly stated that regardless of what technologies or strategies were selected, exploration of the Moon and Mars would be an expensive undertaking. Truly suggested that the technological, educational, and spiritual benefits derived from such an endeavor was worth the cost.

After a brief foray into technical details, Traxler returned to budgetary concerns, questioning the space agency's ability to accurately forecast programmatic costs for long term projects. He asked whether Truly was confident in NASA's estimate for reference approach A ($541 billion). The administrator simply said it was premature to make this determination, but that he believed the program would be "very expensive." Traxler followed-up by asking when NASA would be able to provide firm numbers, to which Truly said it would take three or four years of focused technology development to provide a more definitive estimate. In essence, Truly was suggesting that Congress should invest billions of dollars in technology development programs before the agency could tell it how much the long-term project would cost. During the second day of testimony, with many Congressional concerns presumably confirmed, Chairman Traxler only returned to SEI in an attempt to accurately identify exactly where new money for SEI could be found within the NASA budget request. This was an ominous sign, calling into question whether Congress would provide any funding for implementation of the initiative.[10]

By the end of April, with Congress preparing to eliminate all SEI-related funding from the NASA budget, the Space Council set into motion a concentrated lobbying effort aimed at garnering support for the space program and SEI. The first step

---

[10] House of Representatives, Committee on Appropriations, "Departments of Veterans Affairs and Housing and Urban Development, and Independent Agencies Appropriations for 1991," in *Hearings Before a Subcommittee of the Committee on Appropriations, House of Representatives, One Hundred First Congress, Second Session: Subcommittee on VA, HUD, and Independent Agencies—Part IV: National Aeronautics and Space Administration* (Washington, DC: U.S. Government Printing Office, 1990), pp. 50-57; pp. 136-143.

in the strategy was to hold a "space summit" at the White House. While President Bush met weekly with the senior Congressional leadership to discuss selected subjects, this meeting was notable because it was the first time in American history that space policy would be the sole topic on the agenda.[11] According to an internal White House memorandum, the primary purpose of the summit was for Bush to show support for the FY 1991 space budget. Secondarily, the gathering provided an opportunity to discuss SEI. Not since the initiation of the Apollo program had a president given such high priority to the space program. During the intervening period, space activities were kept alive by a select group of congressional appropriators and top-level NASA officials. The belief within the administration was that this traditional coalition would not be able to deliver on President Bush's ambitious request for a 24% increase in funding for the space agency or obtain approval to implement SEI. Senator Barbara Mikulski and Senator Jake Garn (chair and ranking member of the Appropriation Subcommittee on VA, HUD, and Independent Agencies) had confirmed this opinion, warning the White House that the Moon-Mars initiative was particularly vulnerable in the current budgetary environment, absent strong intervention by the White House. Based on this advice, the White House plan was to have President Bush actively promote the initiative, both publicly and with top congressional powerbrokers.[12]

On 1 May, after being delayed in mid-April by the death of Senator Spark Matsunaga (D-HI),[13] the summit took place at the Old Executive Office Building. The event was attended by sixteen congressional participants[14] and nearly twenty members of the White House staff.[15] President Bush opened the meeting by affirming

---

[11] Andrew Lawler, "Space Summit Set for May: Bush, Quayle Invite Members of Congress to Talk Space," *Space News* (23 April 1990), p. 1.

[12] Mark Albrecht to Fred Mcclure, 16 April 1990, Bush Presidential Records, George Bush Presidential Library; Jim Cicconi to President Bush, 30 April 1990, Bush Presidential Records, George Bush Presidential Library.

[13] In 1986, Senator Matsunaga wrote *The Mars Project, Journey Beyond the Cold War*, an unabashed call for a wide-variety of joint space missions with the Soviet Union and other nations. Matsunaga was one of the U.S. Senate's most outspoken proponents of outer space development.

[14] The congressional participants included: Senate Majority Leader Bob Dole (R-KS), Senator Robert Byrd (D-WV), Senator Mark Hatfield (R-OR), Senator John Danforth (R-MO), Senator Barbara Mikulski (D-MD), Senator Jake Garn (R-UT), Senator Howell Heflin (D-AL), House Speaker Thomas Foley (D-WA), House Majority Leader Richard Gephardt (D-MO), House Minority Whip Newt Gingrich (R-GA), Representative Silvio Conte (R-MA), Representative Robert Traxler (D-MI), Representative Bill Green (R-NY), Representative Robert Roe (D-NJ), and Representative Robert Walker (R-PA).

[15] The White House participants included: Chief of Staff John Sununu, National Security Advisor Brent Scowcroft, Deputy Chief of Staff Andrew Card, Deputy Chief of Staff Jim Cicconi, Communications Director David Demarest, Press Secretary Marlin Fitzwater, NSC Executive Secretary Mark Albrecht, and Chief of Staff to the Vice President Bill Kristol.

his personal commitment to the American space program, which he believed to be of vital importance to the nation's future. He contended that space leadership was crucial to maintaining national leadership in the high tech world—in particular, he lauded the launch of the Hubble Space Telescope (HST) aboard the Shuttle *Discovery* in late April as an example of U.S. accomplishments in space science. He further argued that there were real and tangible benefits derived from investments in the national space program, including revolutions in communications and computerization, advances in industrial materials and medical knowledge, the creation of millions of high-tech jobs, and inspiring future generations of scientists and engineers. President Bush then appealed for congressional support for his increase in civil space spending—which he asserted would put the nation on the path of recovery from many years of underinvestment in space. He made the case that Mission to Planet Earth and SEI embodied what the space program was all about—to use space to examine Earth from above and to push outward to new frontiers. In conclusion to his remarks, Bush acknowledged that Congress was concerned about the proposed investment in the Moon-Mars initiative. To address these issues, he turned the meeting over to Vice President Quayle.[16]

Vice President Quayle began by emphasizing the Space Council's priorities, including: a balanced mix of human and robotic, scientific and exploratory missions; pursuit of challenging initiatives; and pushing space innovation designed to ensure national leadership in cutting edge technology. He then launched into a defense of SEI. Stating that the Council had dedicated significant effort to creating a strategy for SEI, Quayle argued that it was fundamentally in the national interest to implement a new round of exploration that would produce countless direct and indirect benefits. He told the attendees that the council's approach for SEI was to begin a multi-year technology research effort. The administration was asking Congress for the funding ($188 million in FY 1991) and the time to examine alternative ways for better, faster, cheaper, safer ways of reaching the Moon and Mars. Quayle made clear to the assembled congressional leaders that this should not be considered a new program start, but an opportunity to investigate what was involved in achieving the initiative and ultimately to save money. He was adamant on this point, stating unequivocally that the White House was not asking Congress to commit to a new program. Quayle argued, however, that it was important to start the technology research needed to initiate the program immediately, rather than waiting for the program to get bogged down in bipartisan politics during an election year. He also suggested that SEI offered an exceptional opportunity to showcase U.S. leadership during a time of rapid political change around the globe.

---

[16] Talking Points for the President, Congressional Leadership Meeting on Space, 27 April 1990, Bush Presidential Records, George Bush Presidential Library.

Mark Albrecht remembered the congressional position during the summit "hadn't changed much, it pretty much remained the same—highly skeptical." The participants indicated that while they were willing to provide money for studies, they did not believe there was enough justification for a major new program start. Instead, they wanted to see more detail regarding what the actual initiative would look like before they got fully behind the program.[17] One senior congressional aide recalled, "By this time, Chairman Traxler was carrying the message around that 'we can't afford this given our allocation. We can't do it.' He was already negative about it, so coming out of there I don't think he was convinced differently. Congress had already pretty much made up its mind.'"[18]

SEI Hits the Road

In early May, with congressional support still very much in doubt, the Space Council staff began preparing for President Bush to make a major space policy speech. The intention was that this address would provide some much needed focus for the program, and at the same time allay Congressional fears that Bush was committed to a $400 billion, crash program. Mark Albrecht recalled that by this time NASA had "leaked their numbers out to everybody on the Hill, attempting the crib death of this whole initiative. We still didn't have the full support of the space agency, I don't believe even at this time NASA was embracing it. I think they were more worried about the space station than they were interested in setting a new course." To combat this behind the scenes attack, the White House decided that a presidential rebuttal was needed to make it clear that the Administration was not talking about a crash program[19]

On 11 May, ten days after the space summit, President Bush delivered the commencement address at Texas A&I University. He utilized this speech as an opportunity to discuss the role the national space program would play in America's future. Bush told the assembled graduates that SEI formed the cornerstone of his far-reaching plan for investing in America's future, saying, "Thirty years ago, NASA was founded, and the space race began. And 30 years from now I believe man will stand on another planet. And so, I am pleased to return to Texas today to announce a new Age of Exploration, with not only a goal but also a timetable: I believe that before Apollo celebrates the 50th anniversary of its landing on the Moon, the American flag should be planted on Mars." With this speech, Bush set a timetable for SEI and

---

[17] Ibid.; Albrecht interview.

[18] Senior Congressional Aide interview via electronic mail, Washington, DC, 15 December 2004.

[19] Albrecht interview.

answered critics that argued he lacked the vision of a great president.[20] As he had done ten months earlier, he did not speak to the cost of achieving these lofty goals. When asked by reporters as he boarded Air Force One where the money would come from to fund the initiative he simply said, "Thirty years is a long time."[21] Coming in the wake of months of strategizing within the Space Council regarding how to get SEI back on it feet, this answer was most unsatisfying. It left the impression that either Bush was not fully engaged in the decisions that were being made with regard to space policy, or that the White House simply couldn't produce a good answer to this fundamental question.

As had been the case with his speech announcing SEI the previous summer, the reaction to President Bush's commencement address was not entirely positive. *The New York Times* complained he "did not give any estimate…of how much the program would cost. Nor did he discuss whether the mission would be mounted alone or with international partners."[22] *The Washington Post* quoted Senator Al Gore saying, "before the President sets out on his mission to Mars, he should embark on a mission to reality by giving us some even faint indication of where the $500 billion is going to come from."[23] Dick Malow actually felt that setting a 30-year timeframe weakened the initiative on Capitol Hill. It was the "antithesis of the whole Apollo idea. How do you spread an initiative like this over so many presidential administrations?"[24] These reactions from key Democratic leaders pointed to the difficult position the administration still found itself in due to the expensive policy alternative generated within the 90-Day Study. Even some NASA officials felt that setting a timetable for the initiative was a mistake, believing it would drive costs up whereas a 'go-as-you-pay' program would have had a significantly reduced budgetary impact on an annual basis.[25]

Outside the Capitol Beltway, the speech seemed to play even worse. The editorial page of Salem, Oregon's *Statesman Journal* contended, "A rocket trip to Mars begins on a foundation of common purpose and sound finances at home. A nation

---

[20] *Public Papers of the Presidents of the United States*, 11 May 1990, *Remarks at the Texas A&I University Commencement Ceremony in Kingsville, Texas*, http://bushlibrary.tamu.edu/papers/ (accessed 2 January 2003).

[21] Janet Cawley, "Bush Goal: Man on Mars by 2020," *The Chicago Tribune* (12 May 1990); James Gerstenzang, "Bush Sets 2019 for Mars Landing," The Philadelphia Inquirer (12 May 1990).

[22] John Noble Wilford, "Bush Sets Target For Mars Landing: He Seeks to Send Astronauts to Planet by Year 2020," *The New York Times* (12 May 1990).

[23] Kathy Sawyer, "Bush Urges Mars Landing By 2019: Democrats Point to Money Problems," *The Washington Post* (12 May 1990).

[24] Malow interview.

[25] O'Handley interview.

that doesn't know how to balance its budget, reduce a $3 trillion deficit, and fight the decay of its citizens through the effects of drugs, disrupted families and crippled schools will never find the money and willpower to visit the heavens. Bush has given us an empty challenge. We should ask him to return to Texas and repeat the same speech. This time let him add a page at the beginning, one that spells out how this country can first get its feet back on the ground. Then we can head for Mars."[26] The response was no better in *The Seattle Post-Intelligencer*, where the opening line of an opinion editorial read, "Heigh-ho, heigh-ho, it's off to space we go!"[27] A letter the White House received from a local official in Kittery, Maine (just south of President Bush's vacation home in Kennebunkport) suggested, "American pride will best be shown by meeting the needs of all the people here on Earth. $500 billion would make a good start."[28] Once again, this was not the reaction the administration was hoping to elicit from a presidential address on the importance of space exploration.

A few weeks later, Vice President Quayle met briefly with the person the Space Council hoped would be able to build confidence in SEI—Lieutenant General Tom Stafford (USAF-retired). Stafford was a former astronaut who commanded Apollo 10, the 'dress rehearsal' for the first lunar landing mission, and the U.S. portion of the 1975 Apollo-Soyuz Test Project. He had recently agreed to head the Exploration Outreach Program, which had been created by NASA in response to the Space Council request that the agency seek out new technical approaches that might reduce SEI's implementation costs. Under this outreach effort, the space agency expected to obtain wide-ranging ideas through public solicitations, which would be evaluated by the RAND Corporation—a California-based think tank. The most promising of these proposals, and others directly from NASA, DOD, and DOE, would then go to a "Synthesis Group" headed by General Stafford. This group's recommendations would be reviewed by the NRC and reported directly to the Space Council in early 1991. At the meeting with Stafford, Vice President Quayle expressed his belief that the Synthesis Group would serve as a vehicle for generating enthusiasm and support for SEI. He concluded the meeting by conveying his hope that the group would identify at least two fundamentally different approaches to carrying out the initiative.[29]

---

[26] Editorial Board, "Empty Rhetoric Fuels Mars Talk," *The Statesman Journal* (16 May 1990).

[27] Henry Gay, "Reaching For Stars, Er, Mars," *The Seattle Post-Intelligencer* (24 May 1990), sec.A15.

[28] Maria S. Barth to President George Bush, 24 May 1990, Bush Presidential Records, George Bush Presidential Library.

[29] Mark Albrecht to Vice President Dan Quayle, "Meeting with Lt. General Tom Stafford," 31 May 1990, Bush Presidential Records, George Bush Presidential Library; Fact Sheet, "Space

Chapter 5: The Battle to Save SEI

In early June, at the fourth Case for Mars conference, an alternative emerged that would captivate Mars enthusiasts for years to come—and would work its way into NASA planning years later. The most talked about presentation of the symposium was delivered by Martin Marietta aerospace engineers Robert Zubrin and David Baker. Named 'Mars Direct,' their system architecture included several key elements designed to reduce mission costs and increase scientific return, including:

- Direct flight to and from the Martian surface (which eliminated the need to use Space Station Freedom)
- No earth orbit or lunar orbit rendezvous (which eliminated the need for multiple spacecraft)
- Fueling of the Earth Return Vehicle using propellant generated on Mars from the atmosphere
- Extended operations on the Martian surface (up to 555 days)

Although this approach was considered a high risk alternative to the TSG architecture highlighted in the 90-Day Study, Zubrin and Baker argued Mars Direct would only cost $20 billion—approximately one-twentieth the price tag associated with the space agency plan. Because it was based on existing technologies packaged in an innovative system architecture, many in the space policy community viewed this as an option worth serious consideration.[30]

In mid-June, the Bush administration set in motion a flurry of events intended to garner public support for SEI. These activities were commenced largely in response to a House Appropriations subcommittee vote to eliminate all spending associated with the initiative.[31] This lobbying effort began with a series of meetings to brief key actors within the space policy community. Held at the White House, the presentations were tailored to the corresponding audiences in a coordinated effort—with Vice President Quayle and Admiral Truly as the featured speakers. The message conveyed to a group of congressional staffers was that a failure to provide funding for SEI would create the impression that the United States lacked the political will to take risks to expand humanity's reach into the solar system. Reporters that regularly

---

Exploration Initiative Outreach Program," National Aeronautics and Space Administration, 31 May 1990, National Aeronautics and Space Administration Historical Archives; "Chronology of the President's Space Exploration Initiative," National Space Council, Bush Presidential Records, George Bush Presidential Library.

[30] J. Sebastian Sinisi, "Forum Delegates Confident," *Denver Post* (11 June 1990), sec.4B.

[31] James Gerstenzang, "Bush Denounces NASA Fund Cuts," *The Los Angeles Times* (21 June 1990), p. 28.

covered NASA were told that SEI would produce significant economic, technical, and educational benefits for the nation. Finally, industry and academic leaders were informed that SEI would be part of an overarching administration strategy designed to foster innovation by permanently extending the research and experimentation tax credit and reducing regulatory burdens on corporations.[32]

These briefings were immediately followed by a major presidential address on space policy at the Marshall Space Flight Center (MSFC) in Huntsville, Alabama. After attending a fundraising luncheon for Governor Guy Hunt, President Bush arrived at the center for a tour. This included a visit to the Hubble Space Telescope (HST) Orbital Verification Engineering Control Room, where a NASA team was coordinating the adjustment and final checkout of the groundbreaking orbital observatory.[33] Bush then conducted a full press conference on the center grounds. Despite the setting, the majority of the event was spent detailing the American response to ongoing unrest in the Middle East. However, Bush was asked a few questions regarding the national space program, among them one directed at SEI:

Q: A question about space. How serious are you about this lunar base and Mars mission proposal? Would you go so far as to veto the bill that contains NASA appropriations if Congress decides to delete all the money?

A: I haven't even contemplated any veto strategy. I'd like to get what I want. I think it's in the national interest. I think that the United States must remain way out front on science and technology; and this broad program that I've outlined, seed money that I've asked for, should be supported. But I think it's way too early to discuss a veto strategy. We took one on the chops in a House committee the other day, and I've got to turn around now and fight for what I believe.[34]

---

[32] Briefing, "White House Briefings on the Space Exploration Initiative," 8 May 1990, Bush Presidential Records, George Bush Presidential Library; Charles Bacarisse and Sihan Siv to Fred McClure, "Briefing for Key Congressional Staff on NASA's Space Exploration Initiative," 10 May 1990, Bush Presidential Records, George Bush Presidential Library; Charles Bacarisse to Bob Grady, "Briefings for Key Constituent Groups on the Space Exploration Initiative," 5 June 1990, Bush Presidential Records, George Bush Presidential Library; Charles Bacarisse and Sichan Siv to Cece Kramer, 23 May 1990, Bush Presidential Records, George Bush Presidential Library.

[33] "Tour of Hubble Space Telescope Orbital Verification Engineering Control Center," Ede Holiday, 20 June 1990, Bush Presidential Records, George Bush Presidential Library.

[34] *Public Papers of the Presidents of the United States,* 20 June 1990, *The President's News Conference in Huntsville, Alabama,* http://bushlibrary.tamu.edu/papers/ (accessed 2 January 2003).

Chapter 5: The Battle to Save SEI

Following the press conference, the President took the stage before a crowd of 4,000 MSFC employees.[35] President Bush opened his remarks by recalling his campaign speech at MSFC two and a half years earlier, during which he had vowed to launch a dynamic new program of exploration. "I'm pleased to return to Marshall to report that we have made good on these promises," Bush said, "and we've done it the old-fashioned way, done it the American way—

President Bush at MSFC
(NASA History Division, Folder 12601)

step by step, program by program, all adding up to the most ambitious and far-reaching effort since Marshall and Apollo took America to the Moon." He criticized House Democrats for voting to deny funding for SEI-related concept and technology development, stating that partisan politics had led his opponents to turn their backs on progress. He compared them to naysayers in the Court of Queen Isabella who argued against Columbus' voyage that discovered the New World. President Bush stated that during the Apollo era, significant funding for the space program had fostered a golden age of technology and advancement—one that he hoped would be equaled by a permanent return to the Moon and crewed missions to Mars. He concluded his remarks with a challenge for Congress "to step forth with the will that the moment requires. Don't postpone greatness. History tells us what happens to nations that forget how to dream. The American people want us in space. So, let us continue the dream for our students, for ourselves, and for all humankind."[36]

The day after President Bush spoke at MSFC, the administration coordinated a full day of events aimed at further building support for SEI. Newspapers throughout the nation contained opinion editorials written by supporters of the initiative, including: Representative Tom Lewis (R-FL) in the *Orlando Sentinel*; former astro-

---

[35] Mark Albrecht to Jim Cicconi, "Background Materials, June 20 Marshall Space Flight Center Event," 19 June 1990, Bush Presidential Records, George Bush Presidential Library.

[36] *Public Papers of the Presidents of the United States*, 20 June 1990, *Remarks to Employees of the George C. Marshall Space Flight Center in Huntsville, Alabama*, http://bushlibrary.tamu.edu /papers/ (accessed 2 January 2003).

naut Buzz Aldrin in the *Los Angeles Times*; Dartmouth University Professor Robert Jastrow in the *Baltimore Sun*; and former astronaut Eugene Cernan in the *Houston Chronicle*.[37] On Capitol Hill, Representatives Bob Walker and Newt Gingrich hosted a press conference praising Bush for his leadership with regard to the American space program. On the Senate floor, Senator Jake Garn formally introduced the program and Senators Bob Dole, Phil Gramm, and Malcolm Wallop spoke on SEI's behalf. In the late afternoon Vice President Quayle appeared in a series of satellite interviews in targeted states, including California, Florida, Texas, and Virginia. Finally, the Republican National Committee released radio actualities in key districts around the country.[38,39] By mid-June, there was a feeling that "SEI was gaining momentum."[40]

Losing Faith in NASA

Despite any progress that may have been made during mid-June, the emergence of a series of crises at the end of the month halted any momentum the administration had gained. On 26 June, NASA held a press conference to reveal that its engineers had discovered a crippling flaw in the main light gathering mirrors of the $1.5 billion HST. The space agency reported that this defect would mean the largest and most complex civilian orbiting observatory ever launched would not be able to view the depths of space until a permanent correction could be made—which would likely have to wait two to three years for an astronaut visit with newly manufactured parts. Although many of the instruments aboard the HST would still be functional, the impacted wide-field and planetary camera would be inoperable (reducing by 40% the planned scientific work of the platform). Project managers announced they suspected the problem was in one of two precisely ground mirrors, although they were not sure which one. The two mirrors had tested perfectly on Earth, but once in orbit, they failed to perform together as expected—they were not tested together on the ground because of the huge potential expense and inability to replicate a zero-g environment. Associate Administrator for Space Science Dr. Lennard Fisk disclosed that the agency was forming a review board to investigate the problem.[41]

---

[37] Earlier in the week, *Roll Call* had dedicated an entire issue to the space program, with opposing views expressed on SEI from Senator Jake Garn (pro) and Senator Al Gore (con).

[38] A radio actuality is a group of sound bites sent out to radio stations to be used in news reports.

[39] "Moon/Mars Initiative," 19 June 1990, Bush Presidential Records, George Bush Presidential Library.

[40] O'Handley interview.

[41] Warren E. Leary, "Hubble Telescope Loses Large Part of Optical Ability: Most Complex

Chapter 5: The Battle to Save SEI

Two days after the HST revelation, NASA was forced to ground the entire space shuttle fleet because the *Columbia* and *Atlantis* had developed deadly hydrogen leaks. NASA was forced to admit that they did not know the cause of the leaks, although one possibility was a misalignment between the external tank and the orbiter vehicles. Program managers announced that expert teams of engineers were working feverishly to solve the problem before at least two missions were postponed to make way for the upcoming launch of the *Ulysses* spacecraft—a probe designed to study the sun. Coming on the heels of the HST announcement, this effectively killed any energy generated by recent administration activities designed to garner support for SEI.[42] *The Washington Post* opined, "the failure of the telescope, which two months ago rode into space amid great fanfare in the hold of a Space Shuttle, led more than a few Americans to wonder whether their country can get anything right anymore. The questioning became even more poignant…when the National Aeronautics and Space Administration announced that the shuttles, too, would be grounded indefinitely because of vexing and dangerous fuel leakages. [These problems] may foster beliefs that the United States is a sunset power, incapable of repeating its technological feats of the past."[43] On Capitol Hill, Dick Malow recalled thinking that these problems greatly hampered the administration's ability to make a case for SEI. "There were a lot of other things on NASA's plate and that hiccup certainly was a detractor. Given the budget environment, Hubble became the focus and SEI tended to get pushed back" on the congressional agenda.[44]

During this same period, the House Appropriations Committee released its mark-up of NASA's budget. Although the space agency would receive a significant overall increase of $2.1 billion ($800 million less than the president's request), the entire budget for SEI was eliminated. Fears that had been raised during budget hearings in April were confirmed. The committee had surgically removed all new monies associated with the initiative (see Table on next page).

Instrument in Space is Crippled by Flaw in a Mirror," *The New York Times* (27 June 1990), sec.A1; Bob Davis, "NASA Finds Hubble Mirror is Defective," *The Wall Street Journal* (28 June 1990).

[42] Joyce Price, "Chief Calls NASA Funding 'Crucial for U.S. Survival," *The Washington Times* (3 July 1990).

[43] John Burgess, "Can U.S. Get Things Right Anymore? Hubble Telescope, Space Shuttle Problems Raise Questions About American Technology," *The Washington Post* (3 July 1990).

[44] Malow interview.

Space Station (advanced programs)	$20M
Advanced Launch System	$40M
Heavy-Lift Vehicle	$10M
LifeSat/Centrifuge	$8M
Lunar Observer	$15M
Exploration Mission Studies	$37M
Pathfinder Program	$154M
Civil Space Technology Initiative	$45M
TOTAL	$329M

SEI Budget Cuts

The budget report indicated that the Space Shuttle and Space Station programs should remain NASA's top priorities. It stated that even if additional funds became available in the future, they should be directed toward these important programs rather than being targeted at SEI.[45] Chairman Traxler was quoted in the *Washington Post* saying, "We didn't have the money."[46] The Senate Commerce Committee promptly followed suit with the release of an authorization bill that similarly eliminated funding for SEI. Senator Al Gore, who authored the legislation, said he feared that funding the initiative would endanger on-going efforts, in particular the Mission to Planet Earth.[47]

By early July, NASA was clearly reeling from this series of setbacks—with SEI a clear casualty, and in some respects a cause, of the outspoken criticism focused on the agency. The view within the White House was that the space program had been terribly crippled.[48] In his memoirs, Vice President Quayle wrote, "The Shuttle

---

[45] U.S. House of Representatives, 101st Congress, 2d sess., "Report 101-556: Departments of Veterans Affairs and Housing and Urban Development, and Independent Agencies Appropriations Bill, 1991," 26 June 1990; U.S. House of Representatives, 101st Congress, 2d sess., "H.R. 5158," 26 June 1990.

[46] Dan Morgan, "Panel Boosts NASA Funds 17 Percent: Moon-Mars Mission Cut $300 Million, Higher-Priority Items Backed," *The Washington Post* (27 June 1990), sec.A4.

[47] James W. Brosnan, "Senate Panel Cuts Funds for Mars Trip From NASA Budget," *The Commercial Appeal* (28 June 1990), sec.A12.

[48] Albrecht interview.

Chapter 5: The Battle to Save SEI

seemed to be grounded all the time with fuel leaks; the mirror on the Hubble telescope couldn't focus; and the agency was pushing a space station design that was so overblown it looked as if we were asking to launch a big white elephant. The mood at the Space Council was grim.... I was searching for a solution for NASA."[49] On 11 July, Vice President Quayle invited a group of space experts (including Tom Paine, Gene Cernan, Dr. Bruce Murray from the California Institute of Technology, and Dr. Hans Mark of the University of Texas) aboard Air Force Two for a meeting to discuss the systemic problems with NASA. He asked for opinions regarding the appropriate actions, if any, the administration should take. During the meeting, an idea emerged to establish a task force to examine how the space program could be restructured to better support an era of sustained long-term space operations. Quayle liked the idea.

Six days later, Vice President Quayle hosted a second meeting at the White House with senior administration officials Bill Kristol, Mark Albrecht, Admiral Truly, and Chief of Staff Sununu to discuss procedures to create such a panel. Quayle recalled in his memoirs:

> I wanted [the study] to get NASA moving again. If that was going to happen, then the commission had to have the authority to look into every aspect of the space agency. The result was a long negotiation about the commission's scope. Truly's original position was that it should look only at the future management structure of NASA—that is, what would come after the space station was built. "No," I said, "it will look at the current management situation." Truly next tried to exempt programs from review, and I said "No, programs will be reviewed as well." He asked that the space station be "off the table" and said that we would all be better served if both it and the Moon-Mars missions were off limits to the commission. In other words, the commission shouldn't pay attention to all the most important things we were trying to do during the next couple of decades. "I'm sorry," I said to Truly, "but everything is on the table, and let the chips fall where they may...." I tried to soothe Truly's feelings by making the commission report through him to me.

By the end of the meeting, Quayle had directed Truly to put together an outside task force to consider the future long-term direction of the space program. That same afternoon the White House announced the creation of the board. On 25

---

[49] Quayle, *Standing Firm*, p. 184.

July, the White House announced that Martin Marietta CEO Norm Augustine had been selected to chair the Advisory Committee on the Future of the U.S. Space Program. The 12-member task force was charged with reporting its findings within 120 days.[50] For all intents and purposes, combined with the work being conducted by Tom Stafford's Synthesis Group, this put SEI on hold for the foreseeable future.

Perhaps the final blow for SEI came when final House-Senate budget deliberations began a few months later. In early September, upholding the cuts that appropriators had made earlier, the Senate released a bill that eliminated all funding for SEI. The Senate's intent was to maintain NASA's core programs and remove any funding for new projects. The report accompanying the bill explained the Senate's motivation for eliminating support for SEI:

> …the Committee recommendation includes no funds for the Moon-Mars initiative because of the very severe limits imposed upon the overall NASA budget…. The large increase recommended by the President's budget… simply cannot be accommodated within a framework that restrains all domestic discretionary programs at the level insisted upon by the administration…. The Committee believes that it is premature to proceed with an extensive planning and technology program oriented toward a manned mission to return to the Moon and then to Mars without a clear, sustainable revenue source available for such an undertaking. NASA's preliminary studies on a manned mission to the Moon and Mars estimate a total program cost of over $500,000,000,000. The Committee believes that these figures will likely underestimate the potential cost of such a mission, and believe that moving on such an initiative, in the absence of providing a way to pay for it, is ill-advised in an era of enormous fiscal constraint.[51]

---

[50] Mark Albrecht to Vice President Quayle, "Meeting on Air Force II with Space Experts", 10 July 1990, Bush Presidential Records, George Bush Presidential Library; Mark Albrecht to Vice President Quayle, 13 July 1990, Bush Presidential Records, George Bush Presidential Library; Press Release, The White House, Office of the Vice President, 16 July 1990, Bush Presidential Records, George Bush Presidential Library; "White House Considers Inquiry Into NASA: Spokesman Says Panel May Redirect Agency," *The Washington Post* (15 July 1990); Press Release, The White House, Office of the Vice President, 25 July 1990, Bush Presidential Records, George Bush Presidential Library; Mark Albrecht to Arnold Kanter, 25 July 1990, Bush Presidential Records, George Bush Presidential Library.

[51] U.S. Senate, 101st Congress, 2d sess., "Report 101-474: Departments of Veterans Affairs and Housing and Urban Development, and Independent Agencies Appropriations Bill, 1991," 26 September 1990; David Rogers, "Senate Panel Cuts Most New Funding for NASA Project," *The Wall Street Journal* (14 September 1990), sec. A16; Helen Dewar, "Budget Vote Disappoints Space Backers," *The Washington Post* (26 September 1990), sec. A10.

Chapter 5: The Battle to Save SEI

The following month, House-Senate conference report was released, confirming the elimination of all funding for the initiative.[52] The real world result of these budget cuts was the death, at least from a congressional perspective, of SEI.

Dick Malow recalled, "because of the budget crunch SEI was an easy target. By that time, it became viewed as a non-starter. We were barely able to fund Station and were supporting Shuttle strongly. Given the budget climate we couldn't spend $400 billion. The initiative started to fall off the cliff by the middle of 1990. The Administration kept going through the motions, but SEI basically went from birth to death in twelve to fifteen months, and was never heard from again. Station, on the other hand, was continually pushed by the administration."[53] Stephen Kohashi concurred with this assessment, remembering, "the primary concern regarding the FY 1991 NASA appropriation was having to deal with the rising cost profile of previously initiated projects such as the Space Station, Earth Observation System, and space science missions. No one had the stomach to commit to another program start, no matter how modest the initial price given the relative magnitude of out-year costs."[54]

The only good news during this period was the decision by the House and Senate authorization committees to reinstate a small amount of funding for SEI-related research. In middle November, the authorizers approved $21 million for exploration mission studies. The purpose of these analyses was to "seek innovative technologies that will make possible advanced human exploration initiatives, such as the establishment of a lunar base and the succeeding mission to Mars, and provide high yield technology advancements for the national economy…."[55] This would prove to be the only funding the initiative ever received.

The Augustine and Synthesis Group Reports

In early December, with the Advisory Committee on the Future of U.S. Space Policy largely finished with its report, Vice President Quayle held a celebratory dinner at his official residence for the panelists. What was planned to be a salutation for a job well done, however, quickly became a working session that repackaged

---

[52] U.S. House of Representatives, 101st Congress, 2d sess., "Report 101-900: Making Appropriations for the Departments of Veterans Affairs and Housing and Urban Development, and for Sundry Independent Agencies, Commissions, Corporations, and Offices for Fiscal Year Ending September 30, 1991, and for Other Purposes," 18 October 1990.

[53] Ibid.; "High Hopes Plunge Under an Onslaught of Budget Cuts," *The Associated Press* (23 December 1990); Malow interview.

[54] Kohashi interview.

[55] U.S. Senate, 101st Congress, 2d sess., "Public Law 101-611," 16 November 1990.

the group's findings. Over dinner, Quayle asked what exciting new projects the committee was going to propose. Norm Augustine told him that they had ranked five space endeavors in order of priority: space science, technology development, Earth science, creation of a new launch vehicle to replace the space shuttle, and human exploration of Mars. Following a general discussion, OMB Director Richard Darman lectured the group on how budget priorities worked. Listing something last, he said, was an invitation to eliminate it. He asked rhetorically whether the commission really wished to announce to Congress and the public that space exploration was so unimportant that it could be scrapped. As the dessert course was served, the panel members agreed to recast their report—space science would remain the top priority, but the report would articulate the need for a balanced program and would not prioritize the remaining project areas. As Darman left for the evening, his farewell to Quayle was, "Thank you for a fine dinner, Dan. Good thing I came. I saved your damn report."[56]

On 10 December, Norm Augustine presented the findings and recommendations of the advisory panel to the full Space Council. The group's most important finding was that the space program needed to shift its fundamental rationale from one dominated by national prestige, national security, and foreign policy (although these remained contributing motivations) to one predicated on global economic competitiveness and environmental protection. The committee determined that a reinvigorated space program would require increases in the NASA budget of 10% annually, reaching a peak of $30 billion by the end of the decade. This appropriation level would support science and technology programs, Mission to Planet Earth (MTPE), Space Station Freedom, a new heavy-lift launcher, and SEI. The panel concluded, however, that if NASA could not obtain authorization from the Administration and Congress at this level, then MTPE and SEI should be scaled back (if not eliminated).[57] Based on these findings, the advisory committee made five overarching recommendations:

- Sustaining space science programs as the highest priority element of the civil space program
- Obtaining exclusions for a portion of NASA's employees from existing civil service rules or, failing that, beginning a gradual conversion of

---

[56] David S. Broder and Bob Woodward, "When the Vice President is Chairman: Debating Direction of Space Programs," *The Washington Post* (9 January 1992), sec. A16.

[57] *Report of the Advisory Committee on the Future of the U.S. Space Program* (Washington, DC: Government Printing Office, 1990); Mark Albrecht to Vice President Quayle, 7 December 1990, Bush Presidential Records, George Bush Presidential Library.

Chapter 5: The Battle to Save SEI

selected centers to Federally Funded Research and Development Centers affiliated with universities, using as a model the Jet Propulsion Laboratory
- Redesigning Space Station Freedom to lessen complexity and reduce cost
- Pursuing a Mission *from* Planet Earth as a complement to Mission *to* Planet Earth, with the former having Mars exploration as its long-term goal and adopting a go-as-you-pay funding strategy
- Reducing dependence on the Space Shuttle by phasing over to a new unmanned heavy-lift launch vehicle for all but missions requiring human presence.

Although the panel felt it was premature to set a timetable for a crewed mission to the red planet, it did believe that Mars exploration was a valuable long-term goal for the space program because large organizations operate better when they have a challenging objective to guide future planning. The report itself stated that without the existence of an enduring aspiration, "we would lose the jewel represented by the vision of a seemingly unattainable goal, the technologies engendered, and the motivation provided to our nation's scientists and engineers, its laboratories and industries, its students and its citizens."[58]

At a press conference following the Space Council meeting, Vice President Quayle declared his support for the recommendations. He communicated his intention to task the Space Council and OMB staffs to prepare a specific plan for implementing the reports recommendations within 30 days.[59] Quayle wrote later, "The report the Commission submitted…was, no matter how politely phrased and presented, devastating. 'Among the concerns that have most often been heard,' it noted, 'has been the suggestion that the civil space program has gradually become afflicted with some of the same ailments that are found in many other large, mature institutions, particularly those institutions which have no direct and immediate competition to stimulate change.' The Space Council was now the competition, at least when it came to making policy, and we wanted the programs to be cheaper, smaller, and faster."[60] The Space Council staff actually thought the report wasn't hard enough on NASA. Mark Albrecht recalled the findings were "a little milder than we had hoped for and anticipated, but were considered quite strong in the community at large."[61]

---

[58] See note above.

[59] Talking Points, "Augustine Committee Press Conference," 10 December 1990, Bush Presidential Records, George Bush Presidential Library.

[60] Quayle, *Standing Firm*, pp. 185-186.

[61] Albrecht interview.

Press reaction to the Augustine report was relatively muted, mainly because everyone seemed to agree that the committee had assembled a thoughtful collection of recommendations. The Administration was content and Congress seemed to be equally happy. Dick Malow recalled that it was "generally considered to be an excellent study. It was well received, particularly because there was a strong emphasis on science."[62] Representative Bill Nelson, Chairman of the House Space Subcommittee, called it "the report of the decade." Early rumblings within NASA, however, signaled that the agency's engineers and managers were not entirely happy with the report—particularly with regard to the Shuttle. While praising the study in general terms, Admiral Truly expressed his reluctance to condemn the Shuttle. In fact, he urged the administration to consider building a fifth orbiter and to preserve the capability to build additional spacecraft. This signaled a continued disconnect between the strategic direction of the Space Council and NASA.[63]

After the report was released, even outspoken supporters of human spaceflight were saying that it was time to put plans for lunar and Martian exploration on hold. Former NASA Administrator Thomas Paine (a member of the Advisory Committee) was quoted saying "the '90s are going to be a decade of rethinking and regrouping." Ray Williamson from the Office of Technology Assessment concurred, stating "Humans to Mars is much more a question of when than if. But a realization is percolating through the space community that we had better back off. I think the budget issues will force rethinking as to the balance between manned and unmanned missions." Former Director General of the European Space Agency Roy Gibson opined that "we are quite likely to cause European ministerial support for space to decline further if this moment is chosen for a clarion call for a new mammoth [human] space spectacular." The concern in Europe seemed to be that SEI would detract attention from Space Station Freedom, to which the international partners had already committed time and resources.[64]

The following May, with SEI in a basically lifeless state, the Synthesis Group submitted its final report to the White House. Unlike the TSG, the panel focused its architectures on specific strategic goals—which provided policy makers with a somewhat enhanced set of alternatives. Like the TSG, however, the group based all of its options on a singe technical approach. While the former group had focused on

---

[62] Malow interview.

[63] Mark Carreau, "Panel Wants to Phase Out Space Shuttle: White House Backs Changes That Will Transform NASA," *The Houston Chronicle* (11 December 1990), p. 1.

[64] Robert C. Cowen, "Manned Space Programs Ger Message: Throttle Back," *Christian Science Monitor* (19 December 1990), p. 1.

chemical in-space propulsion systems, the latter favored nuclear-thermal systems—this was the fundamental technological difference between the TSG and Synthesis Group. The architectural options introduced by Stafford's panel included:

- **Mars Exploration**: the major objective of this architecture was to conduct scientific exploration of Mars, while the emphasis of activities performed on the Moon was primarily to prepare for Mars missions
- **Science Emphasis for the Moon and Mars**: this architecture's prime focus was balanced scientific return from both the Moon and Mars missions, both robotic and human
- **Moon to Stay and Mars Exploration**: this architecture emphasized permanent human presence on the Moon, combined with Mars exploration, to promote long term human habitation and exploration in space
- **Space Resource Utilization**: this architecture emphasized the development of lunar resources to provide energy for Earth and the production of propellants for lunar and solar system exploration.

The first three architectures progressed in both complexity and resource requirements (although no actual budget estimates were provided), with the third being the closest to the basic reference approach introduced in the 90-Day Study. After receiving the report, the administration chose to evaluate it for a month before releasing it publicly. In early June, 40,000 copies of the colorful 180-page document were circulated to the media, industry, educators, government agencies, and international organizations.[65]

Reaction to the Synthesis Group report was decidedly mixed. Some believed it provided the alternatives the 90-Day Study had not, while others contended it was woefully short on crucial details. Vice President Quayle believed it would serve as a valuable tool in making the case for increased funding for space exploration. George Washington University's John Logsdon, on the other hand, argued the report was "a validation of NASA's argument that there aren't a lot of bright new ideas out there that it hasn't considered." By far the most verbalized criticism of the report was that it provided no cost estimates for the various architectures it introduced. Regardless, the study had little chance of positively impacting the implementation of SEI. By the time the report came out, the initiative was no longer politically viable. The

---

[65] *America at the Threshold: America's Space Exploration Initiative*, Report of the Synthesis Group on American's Space Exploration Initiative," May 1991; David S.F. Portree, *Romance to Reality: Moon and Mars Plans*, available from members.aol.com/dsportree/VH11.htm (accessed 10 February 2003).

White House was not focusing on SEI as a means to reform NASA. Instead, the Space Council was taking initial actions intended to change the space agency's leadership.[66]

SEI Fades Away

In September 1991, two years of White House frustration with Admiral Truly came to a head when NASA Deputy Administrator J.R. Thompson tendered his resignation. The job was a presidential appointment and provided the Space Council with an opportunity to select someone who would support President Bush's vision for the future. Mark Albrecht was responsible for making the selection, but was surprised to find that no one would take the position as long as Admiral Truly remained administrator. Despite being an outspoken critic of the administrator, Albrecht was surprised by how widespread anti-Truly feelings were. After briefing Vice President Quayle regarding the status of the search, he was asked to assess whether there was support for Truly's removal. In early December, Quayle and Albrecht met with three former NASA administrators—Jim Beggs, Thomas Paine, and Jim Fletcher. During the course of the meeting each of the three reiterated a common message—Truly had to go.[67]

After conferring with President Bush, Vice President Quayle summoned Admiral Truly to the White House and requested that he step aside as administrator. He offered to appoint Truly to any open ambassadorship in the world in exchange for his resignation. The administrator said he would consider the proposal. A few days later, however, he sent a message to the Quayle stating he would not resign. "Then he went into utter radio silence for a week, maybe two weeks," remembers a Quayle staffer. Then, out of the blue, Albrecht received a phone call from the newly appointed White House Chief of Staff, Samuel Skinner. Apparently Truly had made an appointment with Skinner, in an attempt to plead his case. Quayle and Albrecht were outraged at the administrator's audacity. It was even more startling, however, when Truly again refused to resign when Skinner reiterated Quayle's earlier resignation request. "I want to hear it from the President's lips," Truly told Skinner. By this

---

[66] Warren E. Leary, "Panel Says Much Research is Needed Now to Reach Mars by 2014," *The New York Times* (12 June 1991), p. 25; Kathy Sawyer, "Build Nuclear-Powered Rocket for Mars Mission, Panel Urges: Experts' Report Offers NASA 'New Approaches,'" *The Washington Post* (12 June 1991), sec. A2; Edwin Chen, "U.S. Mars Visit by 2014, Station on Moon Urged: Presidential Panel Unveils a Controversial Program that Includes Nuclear-Powered Rockets," *The Los Angeles Times* (12 June 1991), p. 1; Paul Hoversten, "Panel Proposes Paths to Moon, Mars," *Gannet News Service* (11 June 1991).

[67] Bryan Burrough, *Dragonfly: NASA and the Crisis Aboard Mir* (New York, NY: HarperCollins Publishers, 1998), pp. 239-243.

Chapter 5: The Battle to Save SEI

time it was early February 1992.[68]

On 10 February, at about five o'clock in the afternoon, Truly was once again summoned to the White House—this time to the Oval Office. After a half hour with President Bush, he finally agreed to submit his resignation. As with most other major space policy decisions made by the Bush administration, there were mixed reactions to the decision to fire Admiral Truly. Many space policy experts were not terribly surprised by the White House move. John Logsdon (a newly appointed member of the Vice President's Space Advisory Board) told *The Washington Post* that Truly "did an extremely valuable job in getting the Shuttles flying again, and restoring a sense of integrity to the agency…[however], Truly's vision of the future was not compatible with the realities of the world." Others were troubled by the signal this forced resignation sent regarding the future course of the space program. Senator Al Gore was quoted saying, "I view this as a very troubling sign that…Quayle's Space Council may have forced Admiral Truly to leave this job because of the [Space Council's] insistence on running NASA from the Vice President's office."[69]

Admiral Truly, President Bush, and J.R. Thompson (Folder 12601, NASA Historical Reference Collection, NASA History Division, Washington, DC)

The day after Truly stepped down, President Bush stopped Mark Albrecht in the hallway as the former was on his way to a meeting. "Your job," the President told him, "is to get me the best NASA Administrator in history, and do it before Truly's resignation is effective." Truly was to resign effective 1 April, which meant that Albrecht only had 45 days to have a replacement in place—which would mean a faster confirmation process than anyone during the entire course of the Bush presidency. Within a few days, Albrecht had compiled a short list of potential can-

---

[68] Ibid.

[69] Richard H. Truly to President George Bush, 10 February 1992, Bush Presidential Records, George Bush Presidential Library; President George Bush to Richard H. Truly, 12 February 1992, Bush Presidential Records, George Bush Presidential Library; Kathy Sawyer, "Truly Fired as NASA Chief, Apparently at Quayle Behest: Ex-Astronaut Feuded With Space Council," *The Washington Post* (13 February 1992), sec. A1; William J. Broad, "NASA Chief Quits in Policy Conflict," *The New York Times* (13 February 1992), sec. A1; Craig Covault, "White House to Restructure Space Program: Truly Fired," *Aviation Week & Space Technology* (17 February 1992), p. 18.

didates. Everyone on the list was well known within the space policy community, except for one name that quickly rose to the top of the heap. Dan Goldin was a relatively obscure middle manager at TRW who a few years earlier had pitched an idea for a smaller, cheaper version of the NASA Earth Observation System (EOS). A mechanical engineer who received his B.S. from City College of New York in 1962, Goldin's first job after graduating was at NASA's Lewis Research Center—where he dreamed of sending humans to Mars. Within five years, however, he left the agency to join TRW and work on classified defense programs. He was a rising star at the company, and in the mid-1980s became heavily engaged in the nation's top-priority Strategic Defense Initiative (SDI). Early in the Bush administration, the National Space Council staff took note of Goldin's dynamic and innovative policies at TRW—particularly his use of very advanced microelectronic technology to launch smaller spacecraft. Albrecht and Vice President Quayle believed Goldin was exactly what the agency needed, someone willing to shake things up and get results. Albrecht recalled having dinner with Goldin and thinking, "this is a keeper, he understands the confluence between technology and risk and cost and schedule." Albrecht became Goldin's biggest champion within the White House. "I always wanted Dan to be the guy," Albrecht remembered, "I kept sending the Vice President lists of names and it always had Dan Goldin on it."[70]

The bigger question the administration faced was whether Goldin, or anyone for that matter, would want to take on the position of NASA administrator. With President Bush's approval ratings down in an election year, anyone who chose to take the position could easily find themselves out of a job if the Democrats retook the White House in November. For Goldin in particular, who had a high paying job in industry, there seemed to be a lot of reasons to stay put in California. Regardless, he was ready for a move and was flattered by the presidential offer. More importantly, he still maintained the love affair with space that he had when he joined NASA in the early 1960s—and he still wanted America to go to Mars. Thus, in early March, he decided to take his chances and agreed to accept the nomination to head the space agency. Just before his nomination was submitted, however, a small problem emerged. In his book *Dragonfly: NASA and the Crisis Aboard Mir*, author Bryan Burrough detailed an astonishing interaction between Goldin and his White House sponsors in early March.

---

[70] Burrough, *Dragonfly: NASA and the Crisis Aboard Mir*, pp. 243-245; W. Henry Lambright, *Transforming Government: Dan Goldin and the Remaking of NASA* (Arlington, VA: PriceWaterhouseCoopers Endowment for the Business of Government, 2001), pp. 14-15; Albrecht interview.

Chapter 5: The Battle to Save SEI

One night Goldin mentioned to Albrecht that, by the way, did it matter that he was a registered Democrat?

Albrecht nearly choked. "Dan, you are to tell no one this," he said. "Do you understand? No one."

Albrecht hung up and phoned Quayle. "I've got fabulous news, he told the Vice President. "Dan Goldin is a registered Democrat."

"You are kidding me."

"No, I'm not."

And then Dan Quayle chuckled and mentioned the obvious. In that case, Goldin ought to sail through his confirmation hearings in the Democrat-controlled Senate.

On 11 March, with this issue settled, the White House announced that it was putting Dan Goldin forward as its nominee to be the next administrator of NASA.[71]

Overall, the response to Dan Goldin after he arrived in Washington and began making the rounds on Capitol Hill was extremely positive. "The general reaction to Goldin," said one backer, "was, 'Jesus, who the hell was that guy? He's great! Where did you find him?'" During his senatorial confirmation hearings the panel greeted him warmly, but cautioned that he was walking into a budget mess. Goldin told the committee he intended to sharpen accountability and control costs in NASA programs in a way that would win more stable funding support in Congress. Responding to concerns that he would simply be a Space Council puppet, Goldin stated, "I will be in charge of NASA." Goldin was approved overwhelmingly with a mandate for change from both the White House and Congress. On the afternoon of 1 April, Goldin was sworn-in during a brief ceremony in the Oval Office.[72] Seven months later, President Bush was defeated for reelection by Arkansas Governor Bill Clinton.

While it was initially unclear where President Clinton stood on space, although

---

[71] Ibid.

[72] Burrough, *Dragonfly: NASA and the Crisis Aboard Mir*, pp.243-245; Lambright, *Transforming Government: Dan Goldin and the Remaking of NASA*, pp. 14-15; Kathy Sawyer, "NASA Nominee Praised at Confirmation Hearing: Committee Members Warn Goldin About Likely Budgetary, Political Problems Ahead," *The Washington Post* (28 March 1992), sec. A6; Ede Holiday to President George Bush, "Swearing-in Ceremony for NASA Administrator Dan S. Goldin", 1 April 1992, Bush Presidential Records, George Bush Presidential Library.

he had supported continuation of space station program during the election, it became obvious early in his tenure that the American space program was not a top priority on his agenda. Within weeks of taking office, he disbanded the National Space Council and tasked Vice President Al Gore with directing national space policy. Gore had been very impressed with Dan Goldin during the latter's confirmation hearings the previous spring, which explains why Goldin was the highest ranking Bush appointee to remain in place under the new administration.[73] In early February 1993, the fate of the American human spaceflight effort became shockingly clear when Goldin was summoned to the White House. During a meeting with OMB Director Leon Panetta, the administrator was informed that President Clinton's budget would cut funding for the space agency by 20%. As a result, there was no alternative but to kill the Space Station program.[74] "The blood drained out of my face," Goldin later remembered. Before the meeting ended, however, Goldin had successfully lobbied for a few days to prepare a working budget that would maintain a commitment to the Space Station. He believed without the Station, NASA had no future—and would certainly never make it to Mars.[75]

Over the subsequent weekend, Goldin summoned key NASA staffers from around the country to a crisis meeting in suburban Virginia. Over the course of a sleepless 72-hours, the team generated three alternatives for shrinking the existing Station plans. The following Monday, Goldin used a collection of Lego building blocks to build primitive models of Plan A and Plan B, and a single cardboard toilet-paper holder for Plan C. That Tuesday, he used the mock-ups at a briefing for President Clinton's senior staff. He was pleasantly surprised at the end of the meeting to get the go-ahead to fully develop the three new options within 90 days in an emergency redesign effort. The space station eventually avoided cancellation, although its budget was slashed by $7 billion over five years. The Clinton administration later brought the Russians into the program as partners on what was renamed the International Space Station—this program became the primary human spaceflight initiative for the remainder of the decade.[76] It was clear that the Clinton administration had no desire to fund human exploration of the Moon and Mars.[77]

Over three years later, in September 1996, the White House National Science and Technology Council released the first comprehensive revision of national space

---

[73] Lambright, *Transforming Government: Dan Goldin and the Remaking of NASA*, p. 17.

[74] Panetta had been trying to cancel the station program for years while serving in the House of Representatives.

[75] Burrough, *Dragonfly: NASA and the Crisis Aboard Mir*, pp. 262-264.

[76] Ibid.

[77] Lambright, *Transforming Government: Dan Goldin and the Remaking of NASA*, p. 20.

policy since the end of the Cold War. The policy stated the United States would maintain a global leadership role by supporting a strong, stable, and balanced national space program. It presented five goals for the space program:

- Enhance knowledge of the Earth, the solar system, and the universe through human and robotic exploration;
- Strengthen and maintain the national security of the United States;
- Enhance the economic competitiveness, and scientific and technical capabilities of the United States;
- Encourage State, local, and private sector investment in, and use of, space technologies;
- Promote international cooperation to further U.S. domestic, national security, and foreign policies.

Explicitly missing from the document was any mention of human exploration beyond Earth orbit. The document simply stated that "the international space station would support future decisions on the feasibility and desirability of conducting further human exploration activities." On a campaign swing through the Pacific Northwest the day after the document was released, President Clinton said the goal of a human mission to Mars early in the next century was too expensive to pursue, and instead affirmed America's commitment to a series of less expensive robotic probes, the first of which was scheduled to land on the planet the following summer. White House Press Secretary Mike McCurry told reporters that ambitions for human exploration of Mars, which would cost upwards of $100 billion, had met with the hard reality of the national budget. "We're not abandoning that concept," McCurry said. "What we believe is that in the era that we're managing our space exploration resources prudently, we ought to establish sufficient grounds for that type of commitment of resources. To commit those kinds of resources now, lacking a scientific basis for that, the President doesn't think is justified." Thus, in the early fall of 1996, human exploration of Mars vanished from the national space policy agenda.[78]

---

[78] The White House, National Science and Technology Council, "Fact Sheet: National Space Policy," 19 September 1996; Brian McGrory, "Clinton Curbs Mars Project: Drops Manned Mission, Backs Robotic Probes," *The Boston Globe* (20 September 1996), sec. A25; Kathy Sawyer, "White House Releasing New National Space Policy: Robots, Not Astronauts, May Travel to Mars," The Washington Post (19 September 1996), sec. A29.

6

SEI, Policy Streams, and Punctuated Equilibrium

"Some say the space program should wait—that we should only go forward once the social problems of today are completely solved. But history proves that attitude is self-defeating...Many an American schoolkid has read the story of Columbus' doubters, and shook their heads in disbelief that these naysayers could have been so shortsighted. We must not let the schoolchildren of the future shake their heads at our behavior."

President George Bush, 20 June 1990

As was discussed in the first chapter of this book, John Kingdon's *Policy Streams Model* and Frank Baumgartner and Bryan Jones's *Punctuated Equilibrium Model* epitomize an innovative approach to analyzing agenda setting and policy formation. Both models were developed because there was a sense that political scientists and policy analysts could benefit from overarching approaches to understanding the policy process. The policy sciences had previously been dominated by a proliferation of theories that dealt with specific policy phases (e.g., agenda setting, adoption, implementation, and evaluation). The goal of the new models was to provide a more comprehensive system to improve policy analysis within large issue areas.[1] *Policy Streams* and *Punctuated Equilibrium* were originally conceived to study social and

---

[1] Parsons, *Public Policy*, pp. 193-207.

economic policy. One of the objectives of this book, however, is to assess whether the models are relevant to the examination of science and technology policy—particularly large space policy initiatives. This evaluation led to the conclusion that the models offer useful methodologies that can be applied to further our understanding of the space policy community and long-term trends in national space policy.

Policy Streams, SEI, and the Space Policy Community

Although the original intention was to primarily utilize the *Policy Streams Model* to guide the historical analysis of SEI's failure, in writing this book it became increasingly clear that the theory could also be used to provide insights regarding important actors within the space policy community. The methodological core of Kingdon's book were hundreds of interviews conducted over four years with congressional staffers, upper-level civil servants, political appointees, presidential staffers, lobbyists, journalists, consultants, academics, and researchers.[2] One objective of these interviews was to determine which players were important in a given policy community.[3] Due to resource limitations for this book, it was not possible to conduct a large number of comprehensive interviews. Instead, a survey was created to identify influential actors within the space policy community. This survey was circulated to a population of civil servants, presidential staff, lobbyists, academics, researchers, and members of industry. Those surveyed were asked to rate the importance of given actors within the community. This direct inquiry allowed for accurate coding of responses into one of four categories: very important, somewhat important, little importance, or no importance. The goals were twofold. First, to determine who the most influential players are within the space arena. Second, to determine whether SEI's policy entrepreneurs effectively engaged key space policy community actors during the initiative's agenda setting and alternative generation processes.

No single actor has the same ability to set the space agenda as the president. Eighty-three percent of those surveyed considered the president very or somewhat important—very important represented 59% (see Table on next page). Although President Bush wasn't heavily involved in SEI's development, he endowed Vice President Quayle with the authority to push forward the initiative. Before ascending to the presidency, Bush was an outspoken proponent of the American space program and was already leaning toward a commitment to human exploration beyond Earth orbit. Armed with a presidential mandate to fashion a long-term strategy, Quayle was able to force SEI onto the national agenda. Throughout the agenda setting pro-

---

[2] These interviews were conducted in an attempt to better understand policy making within two issues areas, health care policy and transportation policy.

[3] Kingdon, *Agendas, Alternatives, and Public Policies*, pp. 231-240.

Chapter 6: SEI, Policy Streams, and Punctuated Equilibrium

cess, he was the crucial policy entrepreneur advocating for the initiative.

Vice President Quayle was aided in this effort by National Space Council Executive Secretary Mark Albrecht. Within the space policy arena, presidential staffers were collectively among the most frequently discussed actors. Ninety-eight percent of those surveyed considered the presidential staff very or somewhat important—very important represented 74%. During the Bush administration, the staff had more influence over the space policy agenda than at any time in the history of the space program. This was primarily due to the broad role of the Space Council. Combined with budget and policy analysts from OMB, OSTP, and the NSC, there was arguably more attention given to space issues than under any other presidential administration. The centrality of Mark Albrecht and Richard Darman as key policy entrepreneurs for SEI provides evidence of staffer influence. Although strong presidential support was a requisite component for pushing the initiative onto the national agenda, having a dedicated staff working on these issues was an important factor.

Presidential appointees were also among the most frequently discussed actors in the agenda setting process. Ninety-six percent of those surveyed considered presidential appointees very or somewhat important—very important represented 81%. As NASA Administrator, Admiral Truly was heavily relied upon to advocate on SEI's behalf. Although several commentators have questioned whether he was ever sincerely supportive of the undertaking, he was critically important in shaping the proposal that was eventually embraced by Vice President Quayle. The full membership of the Space Council added to the influence of political appointees under the Bush administration. With regard to SEI, however, these cabinet-level officials were not heavily engaged during the agenda setting process.

GOVERNMENT ACTORS	*Very Important*	*Somewhat Important*	*Not Very Important*	*Not Important*
President	59%	24%	13%	4%
Presidential Staff	74	24	2	0
Presidential Appointees	81	15	4	0
Civil Servants	28	50	22	0
Congress	36	57	7	0
Congressional Staff	42	50	8	0

Importance of Government Actors in Agenda Setting

Congress is among the most frequently discussed actors within the space policy arena, although it was seen as significantly less important than the administration—the opposite of what Kingdon found when looking at social and economic policy. Ninety-three percent of those surveyed considered Congress very or somewhat important—but very important only represented 36%. Kingdon found that when a presidential administration decides to put forth a proposal, consulting with

Congress can be crucial for the eventual adoption of the initiative—although this doesn't necessarily mean it won't otherwise reach the national agenda. The Bush administration overlooked the importance of conferring with Congress when considering SEI. Although Quayle and Truly did brief key members of Congress, this was only after the initiative had been formulated and the decision had been made to go forward. There was never any effort to create a coalition of congressional supporters or to engage potential congressional policy entrepreneurs. Furthermore, the administration did not adjust any of its plans based on the reactions of the Congress—particularly in the case of important staffers. Within the space sector, these staffers were equally important as members of Congress. Ninety-two percent of those surveyed considered congressional staffers very or somewhat important—very important represented 42%. This significant influence is probably due to the fact that staffers can spend more time gaining an understanding of the relevant technical details. Similar to congressional members, the value of obtaining staffer support was overlooked by key policy entrepreneurs within the administration.[4]

For the most part, non-governmental actors were not considered to be influential in agenda setting within the space policy community. The primary exception was the aerospace industry and interest groups representing those companies, although even this group was considered far less important than government actors. Professional and public interest groups, academics and researchers, mass public opinion, and the media were all seen as having limited significance. The SEI case study appears to be consistent with these general findings. The White House briefed industry leaders before launching the initiative, but the less than enthusiastic response did not sway the administration from its course. Likewise, although several influential academics and researchers were briefed to gauge reaction to the initiative, they did not participate in SEI's development. The media and public opinion played no part in bringing SEI to the national agenda.

NON-GOVERNMENT ACTORS	Very Important	Somewhat Important	Not Very Important	Not Important
Interest Groups: Industry	16%	66%	18%	0%
Interest Groups: Professional	2	38	56	4
Interest Groups: Advocacy	4	27	47	22
Academics & Researchers	2	49	42	7
Media	5	46	40	9
Industry	16	62	18	4
Mass Public Opinion	7	38	50	5

Importance of Non-Government Actors in Agenda Setting

---

[4] Ibid., pp. 34-42.

Chapter 6: SEI, Policy Streams, and Punctuated Equilibrium

The president cannot dominate alternative generation in the same way as agenda setting. This appears to be particularly true within the space arena, where the scientific and technical options require special training and significant effort. Only forty-one percent of those surveyed considered the president very or somewhat important—very important represented 13%. This lack of control over alternatives was one of the most important factors confronting successful adoption and implementation of SEI. Due to a lack of in-house expertise, President Bush and Vice President Quayle were forced to turn to NASA to generate options for the exploration initiative. The agency, however, did not produce actual alternatives. Instead, the agency chose to put forward different timelines for the same basic approach. The lack of internal technical capability meant Bush and Quayle were largely at the mercy of NASA bureaucrats.

GOVERNMENT ACTORS	*Very Important*	*Somewhat Important*	*Not Very Important*	*Not Important*
President	13%	28%	42%	17%
Presidential Staff	49	38	11	2
Presidential Appointees	74	22	4	0
Civil Servants	60	33	7	0
Congress	24	49	27	0
Congressional Staff	33	51	16	0

Importance of Government Actors in Alternative Generation

The presidential staff is considered to be among the most important actors in alternative generation. Eighty-seven percent of those surveyed considered the presidential staff very or somewhat important—very important represented 49% percent. Kingdon contended that the staff has sway because it is able to "engage in the detailed negotiations—with departments, the Hill, and the major interest groups—that will produce the Administration's proposals and that will clarify the Administration's bargaining positions."[5] In the case of SEI, the Space Council actually abdicated its authority to direct the alternative generation process when it allowed NASA to conduct the 90-Day Study in a policy vacuum. Without clear direction from the Council regarding key budgetary constraints, the space agency was allowed to produce a plan that didn't match up with political reality.

Political appointees were mentioned as the most influential actors in the space sector. Ninety-six percent of those surveyed considered presidential appointees (e.g. NASA Administrator) very or somewhat important—very important represented

---

[5] Ibid., pp. 26-27.

74%. Kingdon writes that there are few cases of confrontation between the White House and appointees because the latter "finds it prudent to bend with the presidential wind, and the President finds it politically embarrassing to be portrayed as being at war with his major advisors."[6] SEI was a clear exception to this rule. Admiral Truly clearly wasn't a strong advocate for the initiative. Under his leadership, NASA flouted requests from the Space Council to provide President Bush with a variety of strategic and technical options. Instead, Truly was committed to carrying out a study that drew heavily on technical approaches that had been developed during the past several years by the NASA Office of Exploration—which resulted in the selection of a single program alternative.

Career civil servants have much more influence on alternative generation than on the space agenda. Ninety-three percent of those surveyed considered career civil servants as very or somewhat important—very important represented 60%. Without doubt, NASA civil servants played a crucial role in the alternative generation process for SEI. The TSG was the most important actor in producing options for the initiative—even though it ultimately developed a single highly expensive alternative. The Space Council was highly critical of this result, but was equally to blame because it failed to provide the agency with clear, written guidance adequately explaining the key constraints facing SEI. Combined with Admiral Truly's reluctance to fully support the initiative, this ultimately doomed its chances for adoption and implementation.

Although Congress is a somewhat important actor for alternative generation in the space sector, it is not considered to be among the most influential. Seventy-three percent of those surveyed considered Congress very or somewhat important—but very important only represented 24% percent. As with the agenda, within the space arena it appears that it is crucial to consult with Congress to understand the constraints facing specific program alternatives—as opposed to members of Congress actually developing those options. With regard to SEI, Congress was never brought into this process. Likewise, congressional staffers were not terribly influential in alternative generation. Eighty-four percent of those surveyed considered congressional staffers very or somewhat important—very important represented only 33% percent. Unlike in other issue areas, where staffers become heavily involved in drafting legislation and negotiating agreements between interested parties, most Hill staffers lack the expertise to develop the very detailed technical plans needed for an initiative like SEI. This has often resulted in mission agencies like NASA having an inordinate amount of influence over the alternative generation process.

With a few exceptions, the non-government actors examined do not have a great deal of impact on the alternative generation process. Interest groups representing the

---

[6] Ibid., pp. 27-30.

Chapter 6: SEI, Policy Streams, and Punctuated Equilibrium

NON-GOVERNMENT ACTORS	*Very Important*	*Somewhat Important*	*Not Very Important*	*Not Important*
Interest Groups: Industry	36%	42%	22%	0%
Interest Groups: Professional	45	11	40	4
Interest Groups: Advocacy	9	27	46	18
Academics & Researchers	20	46	28	6
Media	4	20	45	31
Industry	47	36	15	2
Mass Public Opinion	6	19	45	30

Importance of Non-Government Actors in Alternative Generation

aerospace industry were considered only somewhat influential—professional and advocacy groups were much less important. Academics and researchers affect alternatives more than agendas, primarily because they can match the technical expertise enjoyed by civil servants at NASA. For this very reason, the aerospace industry itself was deemed to be the most influential non-government actor. This is not surprising considering the fact that throughout the history of the space program, government contractors have frequently been turned to during the development of technical alternatives. In the case of SEI, non-government actors had no discernible impact on the initial generation of alternatives. During the course of the 90-Day Study, the TSG did not seek inputs from any of the outside actors with the capability to generate strategic and technical options for the initiative. Instead, the agency relied solely on its own expertise and judgment. In the end, the outcome was a process that resulted in the production of a single alternative.

Punctuated Equilibrium, Space Policy, and SEI

The *Punctuated Equilibrium Model* provides a useful tool for better understanding trends within the space policy arena. In particular, it supplies metrics that can be applied to evaluate whether these long-term movements predetermined SEI's fate. A mixture of policy image and venue indicators was drawn upon to conduct this assessment. These results provide a mixed picture regarding the potential for human exploration of Mars to reach the national agenda and obtain support for successful adoption. During the first 30 years of the space age, backing for both the space program and human exploration of Mars fluctuated a great deal within the American public and key institutional venues. Throughout this period, Mars exploration never reached the same levels of support that other projects (e.g., Project Apollo, Space Shuttle, and Space Station) enjoyed. While several indicators suggest there was growing support for the space program in general (and crewed Mars exploration in particular) by the late 1980s, a dramatic 20-year decline in space program budgets called into question the viability of a costly new initiative.

Baumgartner and Jones elected to study media coverage to gauge trends in policy image. The primary information source they utilized was *The Readers' Guide to Periodical Literature*.[7] They coded the number and tone of articles written in a given year to set the context for the agenda process in a given issue area. The first step in this process was to choose the proper keywords to ensure that all of the appropriate articles were included. The results were then entered into a spreadsheet, where they were further coded by different subtopics. One concern was that looking at only one index would not fully capture the nature of public opinion regarding specific issues. Baumgartner and Jones found, however, that "when we compare levels and tone of coverage in the *Readers' Guide* with those of other major news outlets…it makes little difference which index one uses."[8] For this book, the above methodology for monitoring the public agenda was adopted in almost every respect. Articles were tabulated from 1957 to 1996, coding them for topic and tone. Due to the extraordinarily large number of articles examining different aspects of the space program, only articles relating to human exploration were coded. Six broad topics were chosen to compare different exploration areas: Moon, Mars, Space Shuttle, Space Station, Orbital Flight (non-Shuttle or Station), and an "Other" category that encompassed articles relating to other destinations in the solar system or interstellar flight. To code each article's tone, a basic question was asked: would an advocate of an American human exploration program, managed by a civilian sector agency, be happy or unhappy with the title? Over 6,500 articles were coded in this way.

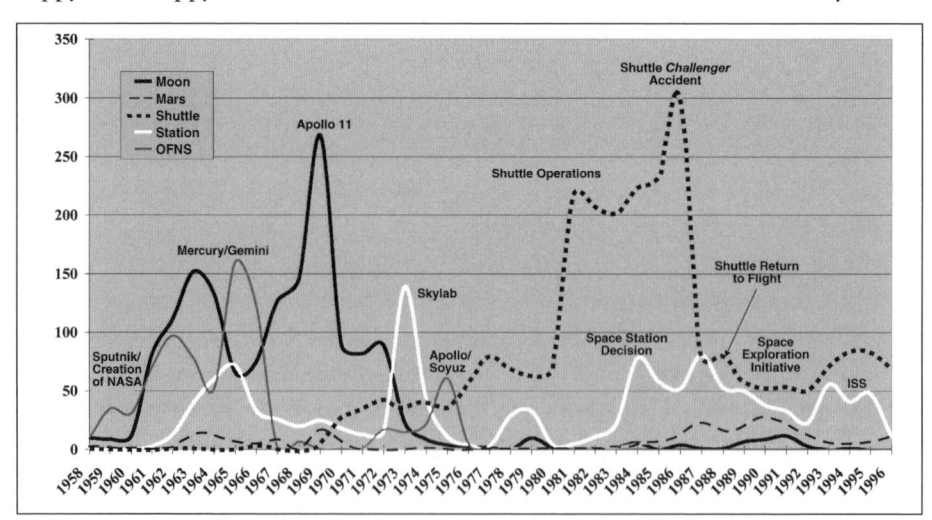

Space Exploration in the Popular Press—By Program (number of articles)
Based on an analysis of the *Readers Guide to Periodical Literature*, 1957-1996

---

[7] The *New York Times Index* was also used to provide supplemental data.

[8] Baumgartner and Jones, *Agendas and Instability in American Politics*, pp. 253-259.

Chapter 6: SEI, Policy Streams, and Punctuated Equilibrium

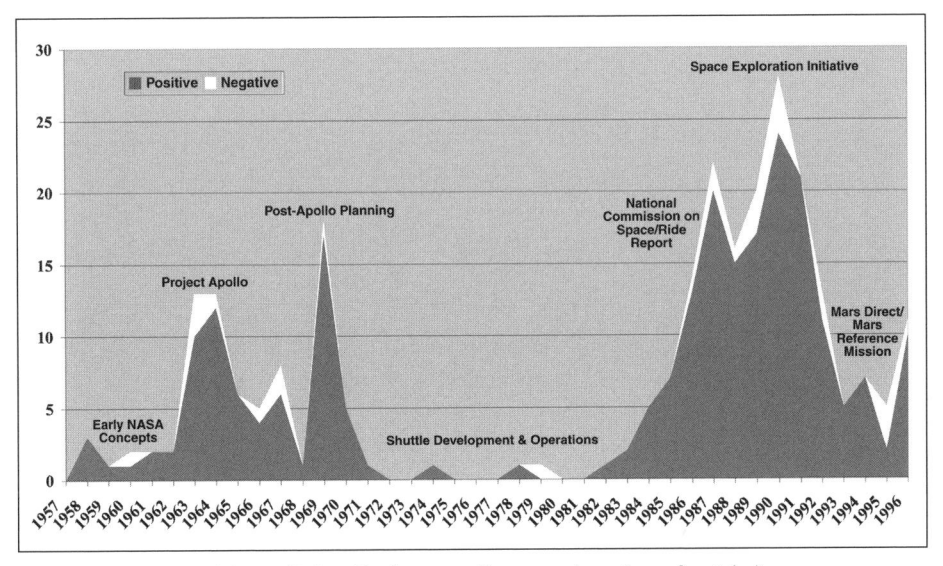

Tone of Mars Exploration Coverage (number of articles)
Based on an analysis of the *Readers Guide to Periodical Literature,* 1957-1996

During the early years of the space age, while the popular press was clearly focused on NASA's efforts to orbit the first American and send humans to the Moon, Mars exploration was largely overlooked. Media coverage for crewed missions to the red planet was trivial compared to other space efforts. Although there were only a small number of articles, media coverage of Mars exploration during the Apollo era was largely supportive, with nearly 90% of the items positive in tone. Coverage slowly increased and reached an initial peak of nearly 20 articles in 1969, when post-Apollo planning was taking place within the federal government. By the 1970s, in the aftermath of the failed effort to push crewed missions to Mars onto the national agenda, media coverage plummeted to at most one article every couple of years. During this period, the majority of media attention focused on Space Shuttle development, the Skylab space station, and the Apollo-Soyuz Test Flight Program. Over the course of the 1980s, media coverage of Mars exploration began growing again, although it was still dwarfed by the Shuttle and Space Station Freedom programs. This increase was the result of "softening up" activities like the Case for Mars conferences, the National Commission on Space, and the Ride Report. By the end of the decade, as SEI reached the national agenda, reporting of crewed Mars exploration represented a significant portion of total coverage. While the reporting was highly supportive, there was no massive increase in attention as there had been for Project Apollo. Regardless, coverage of Mars exploration reached all time highs after President Bush announced SEI, with over 20 articles written annually. In the aftermath of the initiative's failure, however, media attention plummeted. By the mid-1990s, fewer than ten articles were being written per year as the Clinton administration focused on completion of the International Space Station.

Baumgartner and Jones found that survey research, if available in a systematic form, is another data resource for observing changes in policy image. Survey research has become a progressively more important tool for policy analysts and policy makers.[9] The polling data utilized for this book was compiled by the NASA History Division and reveals interesting trends in mass public opinions regarding the American space program. During the post-Apollo planning period, the vast majority of the American public believed that the NASA budget was too large. This provides a compelling reason for the failure of human exploration of Mars to garner support from President Nixon. Over the next 20 years, however, this attitude toward the NASA budget shifted significantly. In the years leading up to the announcement of SEI, there was a relatively high level of support for the space program. By the late 1980s, the majority of Americans believed that NASA spending was either just right or needed to be increased. In 1989, the year that SEI was announced, that figure reached sixty percent for only the fifth time in the post-Apollo era. Despite this relatively high level of public support for NASA, human exploration of Mars was still seen as a lesser priority. Robotic exploration of the solar system and construction of a crewed space station consistently received more support from the general public. Regardless, when the initiative was announced, more than 60% of the public supported establishment of a human outpost on Mars.

As SEI began experiencing difficulties and NASA was dealing with problems in the Hubble and Shuttle programs, however, those numbers began dropping once again—falling to below 50% by December 1990. Although these numbers would eventually recover slightly, by that time the policy window for Mars exploration had already closed.

---

[9] Polls are designed to provide interested parties with information estimating how the mass public would respond to specific closed-ended options, such as "should the government fund human trips to Mars?" The fundamental principle of polling is that a sample population can represent the entire population if there is a sufficient sample size and the chosen methodology ensures a randomly selected sample. Opinion polls are utilized to measure: values, basic beliefs held by individuals that are relatively immune to change and perform a vital role in individuals' lives and choices; opinions, judgments about current issues and policy alternatives; and attitudes, a category between values and opinions representing well thought out views utilized to evaluate new issues and alternatives. While polls measuring values and attitudes are useful because they provide information regarding long-term beliefs, the vast majority of polls relevant to policy makers assess opinions regarding contemporary policy issues. [Mathew Mendelsohn and Jason Brent, "Understanding Polling Methodology," *ISUMA* (Autumn 2001): pp. 131-136.]

Chapter 6: SEI, Policy Streams, and Punctuated Equilibrium

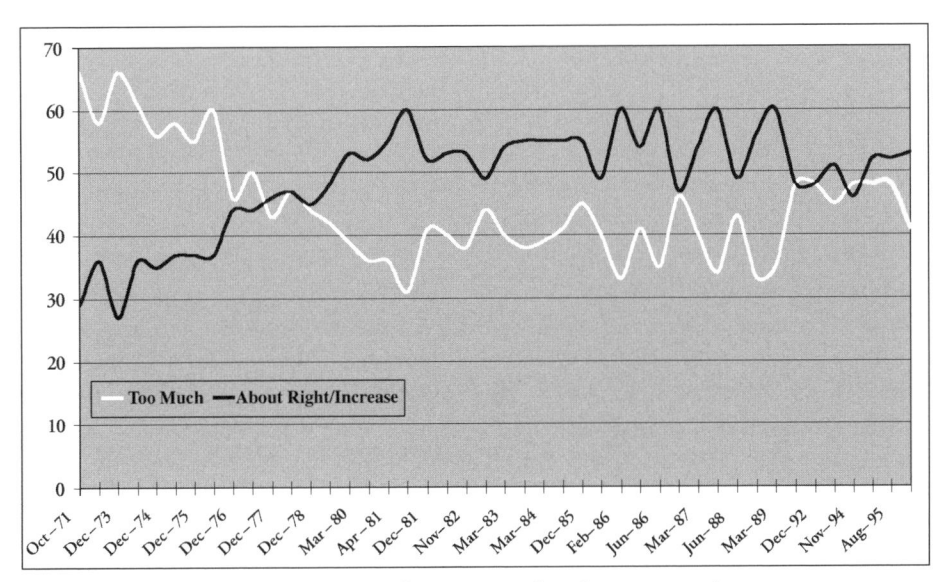

Public Opinion of NASA Spending (as percentage)

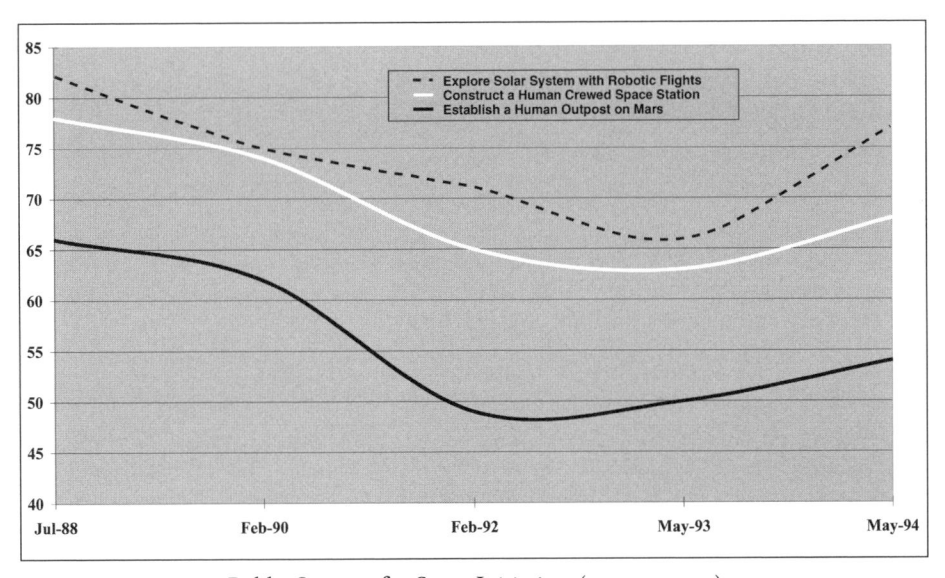

Public Support for Space Initiatives (as percentage)

While examining media coverage and opinion polls provides information relating to policy image, other measures are needed to evaluate venue access. Four different indicators were selected for this book to monitor changes in venue access for space policy issues. The first indicator relied upon an analysis of the Congressional Information Service Abstracts (CIS annual)—a yearly compilation of data on Congressional hearings. In the late 1990s, Baumgartner and Jones established the *Center for American Politics and Public Policy* to provide researchers with tools to better understand the dynamics of policy change. Under the auspices of the center, more than 67,000 congressional hearings were classified using a common policy content code to ensure compatibility over time.[10] For the purposes of this book, the entire CIS annual dataset was downloaded. A separate dataset was created, which included only hearings that related to the American space program. This dataset included over 550 hearings covering the years from 1958 to 1994.[11] The second indicator utilized to monitor changes in venue access, which will be discussed in concert with the CIS annual, relied upon an analysis of the *Public Papers of the Presidents*.[12] For this study, a methodology similar to that used for coding the *Readers' Guide* was employed. Presidential addresses and speeches were coded by topic from 1957 to 1996. This included all papers relating to the civilian space program. These papers were then used to assess trends in venue access for agenda items relating to the American space program.

During the 1960s, presidential support for the nation's space program reached its apex. President Kennedy's decision to send humans to the Moon initiated a decade of close White House attention.[13] Through the '60s, Presidents Kennedy, Johnson, and Nixon delivered well over 400 addresses and speeches relating to the civilian

---

[10] Baumgartner and Jones found that the new CD-ROM format of the CIS annual permitted for efficient searches of policy issues—primarily because hearings are cross-referenced, reducing the possibility of double counting specific hearings, and streamlined selection of keywords. The dataset that was produced included a wide variety of information including the year of the hearing, the committee holding the hearing, and a summary of the topics discussed.

[11] Bryan Jones, John Wilkerson, and Frank Baumgartner, "Policy Agendas Project," *Center for American Politics and Public Policy*, available from depts.washington.edu/ampol/navresearch/agendasproject.shtml; (accessed 18 December 2001.)

[12] In 1957, the *Public Papers* series was created to provide an official compilation of Presidential letters, addresses, speeches, proclamations, executive orders, and other publicly issued materials. Volumes dealing with the Hoover, Truman, Eisenhower, Kennedy, Johnson, Nixon, Ford, Carter, Reagan, Bush, and Clinton administrations are incorporated into the series. In addition, the Clinton papers are now available online and there are plans to expand online coverage to include all presidential papers. The information included in the *Public Papers* is indexed in two ways: by subject and by name. The index entry for each subject contains a word or phrase that identifies the topic and one or more page numbers.

[13] Measured by tabulating the number of space-related addresses and speeches delivered (and catalogued in the *Public Papers of the President* series) by the president in a given year.

Chapter 6: SEI, Policy Streams, and Punctuated Equilibrium

space program. At the height of interest in Project Apollo, presidential addresses and speeches frequently topped 40 per year—and reached as high as 70. This indicated significant access to an important policy venue (e.g. The White House). Similarly, Congressional interest in space exploration peaked during this period.[14] In particular, there was a pronounced spike in congressional interest following the Apollo 1 fire that killed three astronauts (see Figure on page 152).

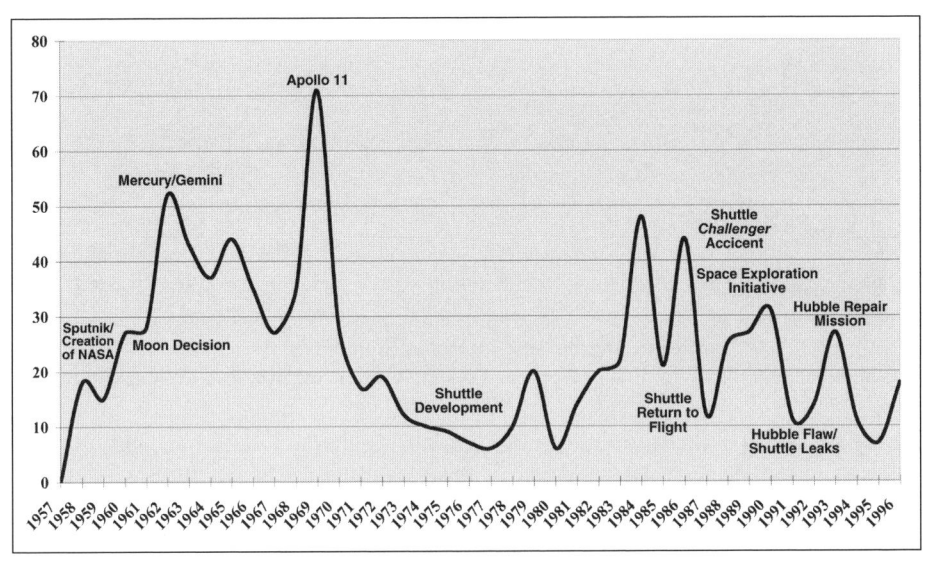

Trends in Presidential Attention (number of speeches/addresses)
Based on an analysis of the *Public Papers of the Presidents of the United States*, 1958-1996

---

[14] Measured by tabulating the number of space-related Congressional hearings held (and catalogued by the Congressional Information Service) in a given year.

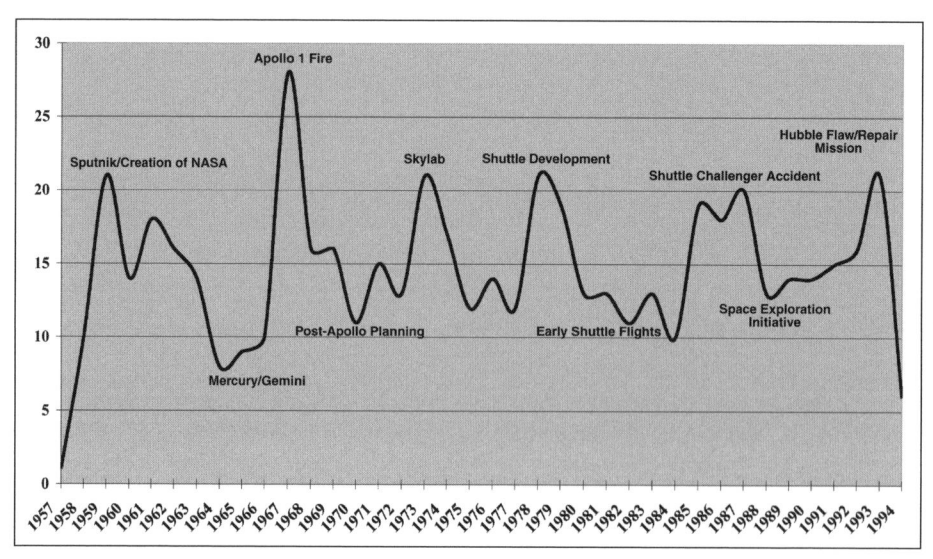

Trends in Congressional Attention (number of hearings)
Based on an analysis of the 'Policy Agendas Project,' Center for American Politics and Policy

By the mid-1970s, the number of presidential speeches and addresses had plunged below ten annually as NASA concentrated on building the Space Shuttle and sending robotic probes to explore the solar system. While Congressional attention dipped below Apollo era levels as well, the number of space-related hearings remained relatively stable at about 15 a year. This was primarily due to annual appropriation and authorization hearings. As the Space Shuttle began operations during the 1980s, presidential addresses and speeches began slowly rising again—with increased interest surrounding the decision to build the Space Station and in the aftermath of the *Challenger* accident. Although Congressional interest remained steady during this period, there were noticeable spikes in interest surrounding the first Space Shuttle flight and as Congress held hearings following *Challenger*.

While presidential interest in the space program was relatively high during the SEI era, congressional support was not particularly robust. President Bush was clearly interested in the space program, as evidenced by his espousal of human missions to the Moon and Mars. During his presidency, he made a number of significant space policy speeches, most directly related to gaining support for SEI. Still, compared with presidential attention during the first decade of spaceflight and even during the Reagan administration, Bush made fewer annual speeches and addresses relating to the space program. On its face, it would appear that Mars exploration had only modest access to this important policy venue during the Bush presidency. This discounts, however, the important role played by Vice President Quayle and the Space Council staff during this period. Combining the involvement of Bush and Quayle with a dedicated internal policy staff to work on space issues, this admin-

Chapter 6: SEI, Policy Streams, and Punctuated Equilibrium

istration was probably more engaged in this arena than any other during the post-Apollo period. In contrast, Congressional interest was relatively low at this time. After a peak following the loss of the *Challenger*, Congressional attention waned as the Bush administration was pushing for SEI. The next peak did not occur until the problems with the Hubble and Space Shuttle programs came to the fore. This indicates that during the post-Apollo period, Congress has been highly reactive to problems within the space program. At the same time, it has not been terribly engaged during relatively calm periods. The lack of access to this critical venue was likely a contributing factor in the eventual failure of SEI.

A third indicator utilized to monitor venue access relied upon an analysis of technical strategies for human exploration of the Moon and Mars. This metric provided insight into Mars exploration's ability to garner attention within a key policy venue—the federal bureaucracy (e.g. NASA). In 1996, NASA Johnson Space Center Historian David Portree created a website called *Romance to Reality: Moon and Mars Plans*. It was a comprehensive catalog of classic, seminal, and illustrative human exploration plans. The majority of these studies were conducted under the auspices of government agencies and private sector companies.[15] The site included detailed summaries and descriptions of reports dating back more than five decades.[16] Portree emphasized studies that emerged as important to later mission planning, but also included reports that helped illustrate essential strategic architectures. For this book, the above reports were used to observe trends, both inside and outside the federal government, in the generation of Moon-Mars exploration plans. Each report from 1950 to 1996 was coded using the same three categories Portree employed: Moon plan, Mars plan, and Moon/Mars plan. Nearly 300 technical studies were coded in this way.[17]

During the course of the 1950s, eight major studies were conducted by government agencies and aerospace companies as part of the "softening up" process for Mars exploration. The following decade, as NASA was working full throttle to meet President Kennedy's lunar landing deadline, more than 50 Mars exploration studies were conducted in preparation for a post-Apollo space program. Despite its growing

---

[15] This database also included a few studies conducted by national commissions (e.g. National Commission on Space), academic institutions, and interest groups.

[16] Portree's goals for the *Romance to Reality* site were fourfold: to educate interested parties about the challenges and opportunities of exploring the Moon and Mars; to make the ideas of engineers and scientists widely available to the mass public; to provide an exciting glimpse of possible futures by looking into the past; and to foster the construction of a future that includes human activity on both the moon and Mars.

[17] Since this analysis was conducted, Portree has changed the format of the *Romance to Reality: Moon and Mars Plans* website. Thus, my results are based on data taken from the website as it existed in March 2002.

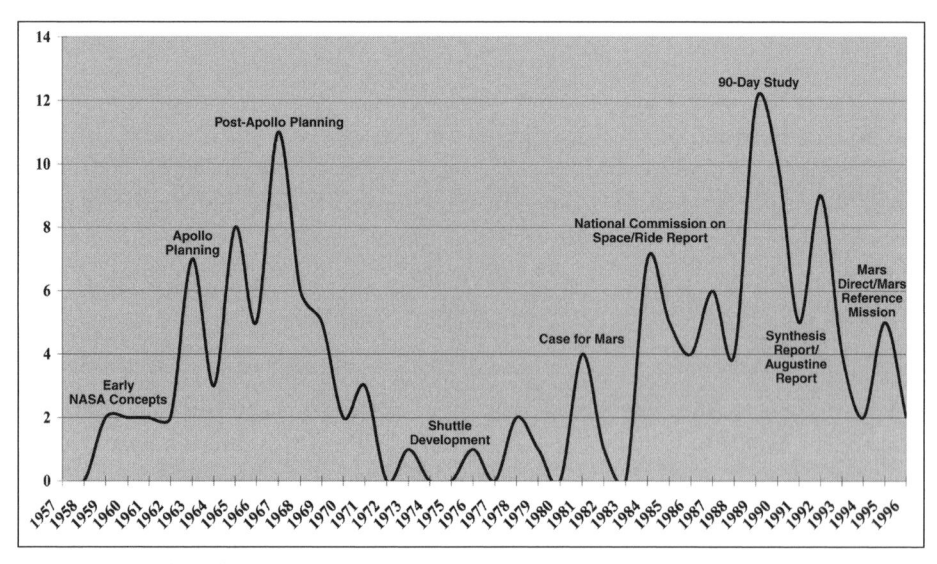

Technical Reports Focusing on Mars Exploration (number of studies)
Based on an analysis of *Romance to Reality: Moon and Mars Plans,* 1957-1996

profile within the space agency, however, Mars exploration had relatively low visibility with the American public. Likewise, it was not supported by President Nixon or Congress—which led to its exclusion from post-Apollo planning. In the 1970s, NASA's official interest in crewed exploration of the red planet basically disappeared. An illustration of this fact is that the space agency did not conduct one major Mars-related study from 1972 to 1985. At the same time, however, private actors kept the dream alive by generating over 20 reports. These were partially responsible for the Reagan administration's decision to place human exploration beyond Earth orbit on the national agenda. By the time President Bush announced SEI, Mars exploration had greater access to the bureaucratic venue than at any other time in the first 40 years of spaceflight. During the four years of the Bush administration, over 35 different studies were conducted. The dilemma for the Space Council, however, was that the politically infeasible 90-Day Study became so closely associated with the initiative.

While the three preceding venue access indicators suggest that by the late-1980s there were favorable trends supporting a major new human spaceflight initiative, a final indicator paints a very different picture—the federal budget. At the height of Project Apollo, NASA's budget was $6 billion, or about 4.5% of the entire federal budget. This represented an extremely large financial commitment to the American space program, which was allocated primarily in an effort to beat the Soviets to the Moon. By the time Neil Armstrong and Buzz Aldrin landed in the Sea of Tranquility, NASA's budget had already begun a rapid decline. Although Congress remained

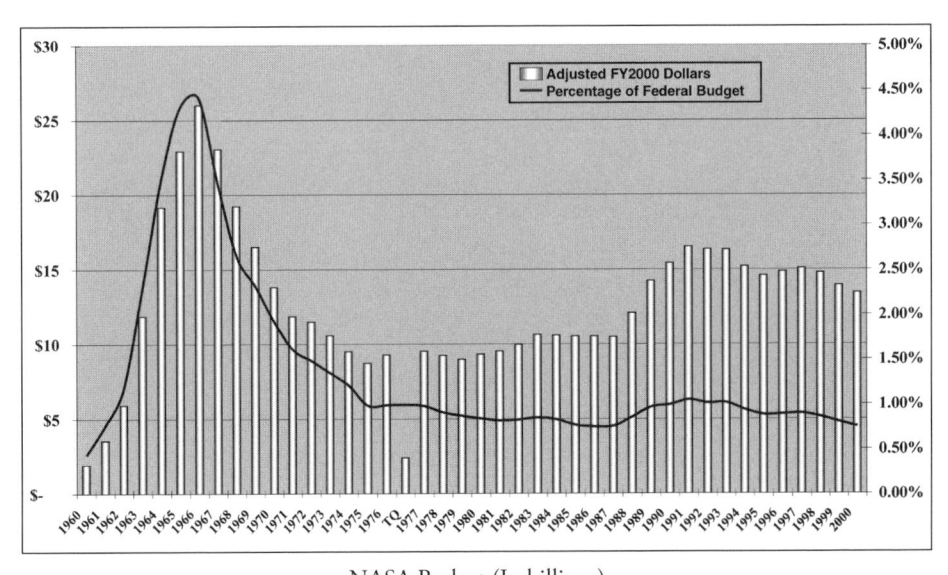

Chapter 6: SEI, Policy Streams, and Punctuated Equilibrium

NASA Budget (In billions)
OMB, "Budget of the United States Government, Fiscal Year 2004: Historical Tables"

engaged in the space program during the next two decades, it was nevertheless steadily cutting NASA's budget from its Apollo peak. The mid-1970s saw a budgetary low of $3.25 billion, with budgets increasing slightly into the early 1980s. More significantly, the NASA budget dropped to under 1% of the entire federal budget—where it would remain permanently except for a three-year period in the early 1990s. Despite the fact that it had experienced steadily decreasing resources, NASA continued to conduct its program planning by assuming that it would receive future budget increases. This tendency to overcommit itself did not foretell a positive result for NASA-led development of another major exploration initiative. While the Bush administration increased funding for NASA, the available resources were far below those available during the mid-1960s. There was little public or congressional support for dramatically increasing the NASA budget. This did not bode well for SEI, which as envisioned by the TSG would have required doubling or tripling the annual allocation for NASA.

Baumgartner and Jones's approach to studying agenda change provides an interesting perspective for understanding larger themes within the American space program. Using policy image and venue access indicators not only reveals interesting trends, it provides insight into factors that helped SEI reach the national agenda but dramatically reduced its chances of gaining Congressional support. As discussed above, there have been striking shifts in media coverage over the past four decades. During the nine years after President Kennedy announced the Moon decision, over

153

2,100 human spaceflight-related articles were written (an average of 235 annually). During the subsequent nine years, as NASA was developing the Shuttle, under 1,200 articles were written (125 annually). The ten years after the Shuttle began operations, including those years following the *Challenger* accident, saw another upward shift with nearly 2,200 articles written (220 annually). This included increased coverage of Mars exploration, which was largely positive in tone. Finally, during the eight years after President Bush announced SEI, coverage plummeted to fewer than 1,000 articles (110 annually). This trend suggests that there have been relatively extended periods of general excitement about the space program within the general public (which resulted in significant media coverage), but that these periods have been followed by equally long periods where the public becomes disengaged. These declines in overall media attention seem to be correlated with periods of poor economic performance and tightening federal budgets. SEI came to the fore toward the end of a cycle of increased media coverage, which enhanced the likelihood it would successfully reach the national agenda. At the same time, however, an examination of past polling data reveals that economic forces were working against any dramatic increase in NASA's annual appropriation. In the end, these budgetary pressures were far more important than any perceived public support for Mars exploration.

An analysis of the *Public Papers of the Presidents* and the CIS annual reveals that the White House and Congress are largely reactive when it comes to addressing space policy issues. Past presidents have delivered the majority of their speeches in reaction to programmatic successes and failures. Presidents Kennedy, Johnson, and Nixon gave regular speeches during the triumphant Mercury, Gemini, and Apollo programs, while President Reagan spoke out primarily in the aftermath of the *Challenger* accident.[18] The fact that President Bush delivered a relatively large number of space-related speeches during his tenure suggests that the space program had significant White House access during this period. This is particularly true because there were no great successes or failures of a similar magnitude during Bush's presidency. Combined with Vice President Quayle's heavy involvement in space policy making, this represented one of the most active administrations in this issue area. Access to this important institutional venue made possible the elevation of SEI to the national agenda. The relative lack of congressional interest, however, tremendously reduced the initiative's chances of actual adoption. The space program has enjoyed relatively stable congressional attention, with significant peaks after major failures (i.e., Apollo 1, *Challenger,* Hubble). When SEI was being considered, there was a clear lull in congressional interest as the legislature focused on an imposing

---

[18] President Reagan's 1984 State of the Union, where he announced his decision to approve the Space Station Program, was the clear exception.

Chapter 6: SEI, Policy Streams, and Punctuated Equilibrium

budget crisis. While this did not necessarily preclude adoption of any new human spaceflight program, it did indicate that promotion of projects requiring large budgetary increases was ill-conceived.

An examination of past technical reports shows that there were two clear periods of interest in Mars exploration, one leading up to post-Apollo planning and another leading up to the announcement of SEI. Prior to both attempts to garner support for adoption of such a program, NASA and non-government actors conducted a steadily increasing number of studies that provided the technical background for the eventual policy alternatives that were considered. Particularly in the case of SEI, this indicates that significant bureaucratic forces were aligned to force an aggressive exploration program onto the national agenda. As discussed above, the fiscal constraints quashed these plans. The federal budget is the single most effective indicator for evaluating the potential for bringing about dramatic programmatic change at NASA. Budgetary trends expose Project Apollo as a clear outlier in the history of the space program, where a unique political environment led to the program's adoption. Subsequent experience has proven that without the emergence of a similar crisis environment, the space program will not receive a large infusion of resources to carry out aggressive human spaceflight programs. SEI's failure is the quintessential example of this lesson. Regardless of bureaucratic desires, technical plans for human exploration beyond Earth orbit must be fiscally feasible.

The policy image and venue access indicators utilized for this study provide a relatively consistent picture regarding the potential for SEI to reach the national agenda and to be successfully adopted. With regard to agenda setting, a combination of metrics suggests that Mars exploration would receive favorable consideration as the long-term goal of the human spaceflight program. These included increased media coverage of Mars exploration, general public support for the establishment of a Martian outpost, and a growing number of reports providing the technical details for such an undertaking. Combined with strong support from the Bush White House, this virtually guaranteed that the initiative would be pushed onto the national agenda. With regard to actual adoption, however, a number of other indicators suggest that SEI faced an uphill battle. Most important among these were fiscal constraints, limited public support for increased NASA budgets, and no congressional backing for expensive new programs. While a less costly Mars exploration program may have been able to gain approval under these circumstances, after the release of the 90-Day Study, the ultimate failure of SEI was assured.

While the above analysis provides some evidence that both the *Policy Streams Model* and *Punctuated Equilibrium Model* provide valuable insight into the rise and fall of SEI, a concluding discussion is in order to generalize these findings beyond this specific case study. This is logically assessed by answering a simple question: Do we know something about SEI we wouldn't have without using these two models? There are at least three potential answers—a lot, a little, or nothing at all. It seems like the most defensible answer lies in the middle. To start with, although we may instinctively have a sense of who the important actors are within a given policy community, Kingdon provides an effective framework for quantifying these beliefs. The survey used for this book was designed to determine whether the appropriate actors were involved in the policy process for SEI. While we may have come to the same conclusion even if we didn't have the survey results, they provide some actual data to back up our assumptions. This survey, or some instrument like it, could just as easily be used to better understand the policy community for other science and technology issue areas. More importantly, it has the ability to provide policy makers with a real world tool for deciding who to engage during the agenda setting and alternative generation processes.

The operational indicators introduced by Baumgartner and Jones, and those developed specifically for this book, also provide a potentially valuable tool for science and technology researchers. Most of these indicators provide good data for large issue areas. For example, the data collected for this book provides revealing trends for space policy as a whole. The data is not as good, however, when one drills down to the next level of detail. Public opinion polls, presidential speeches, and congressional hearings can only be used to gauge large scale trends within an entire field. They cannot easily be used to examine specific issues within this area, such as interest in Mars exploration or space science or orbiting space stations. Although it requires a good amount of effort, it is possible to use data sources like media coverage and technical studies to gain some appreciation of interest in these more specific issue areas. Overall, however, these indicators are a bit cumbersome to use—although this would be made easier if this data, regarding space policy issues in particular, were readily available to policy makers. That is not currently the case, which makes the use of this set of indicators somewhat impractical in the real world.

Therefore, the *Policy Streams Model* and *Punctuated Equilibrium Model* have provided an understanding of the failure of SEI that the case study alone may not have provided. We have better insight regarding who the important actors are within the space policy community, which reveals weaknesses in the agenda setting and alternative generation processes for SEI. We have a better insight regarding larger trends within the space policy arena (i.e., public opinion, congressional attention, federal budgets) that conspired against the adoption of this costly human spaceflight initiative. This type of data not only informs an academic work like this book, but can (and probably should) be used by policy makers in the real world. There does not seem to be any reason why these methodologies could not be applied widely within the science and technology policy field. While this would clearly require a good amount of work, both compiling the data and periodically updating it, the potential benefits for successful agenda setting and alternative generation processes would be worth the effort.

7

The Lessons of SEI

"The story of the dreams and the unbuilt spaceships for flights to Mars should be recorded so that in the future people can examine past ideas of space travel just as we can examine the unconsummated ideas of Leonardo da Vinci by reading his notebooks. Years from now people should be able to decide for themselves whether they want to go to Mars or if they prefer to stay earthbound. But let us not destroy the dream, simply because we do not wish to pursue it ourselves."

NASA Historian Edward Ezell, 1979

During the agenda setting and alternative generation processes for SEI, key policy entrepreneurs did not adequately heed the lessons of the past—particularly those learned from the unsuccessful attempt to place Mars exploration on the government agenda during post-Apollo planning. At that time, the Space Task Group and NASA failed to account for contemporary fiscal and political constraints. This led to the STG endorsement of a Mars exploration approach that required doubling the space agency's annual budget. This was contrary to President Nixon's philosophy and the budgetary environment, which resulted in the eventual failure of the initiative to reach the government agenda. In 1989, exploration of the Moon and Mars gained vital support from President Bush as his administration sought to provide direction to a directionless agency. The policy process that the Space Council nominally directed, however, failed to provide adequate guidance regarding the constraints confronting adoption of the initiative. As a result, NASA's 90-Day Study was significantly at variance with what Congress judged to be in the long-term interest of the nation. Current policy makers are facing similar issues.[1] Future policy makers will surely face them as new policy windows open, providing opportunities to shape the national space program. Understanding the lessons of SEI provides a chance to avoid sharing SEI's fate.

One of the primary causes of SEI's failure was a lack of clear policy guidance from the White House. This deficiency began in early spring 1989, when announcing a robust human exploration initiative was first contemplated by the Bush administration. While Vice President Quayle and Mark Albrecht clearly believed such an undertaking would provide NASA with needed direction, the administration did not have well developed substantive ideas for future programs. Consequently, the Space Council relied heavily on the space agency to provide the details necessary to make an informed decision regarding the technical feasibility of human exploration beyond Earth orbit. The problem, however, was that adequate instructions were not provided regarding the key constraints that should guide the development of a programmatic approach. As a result, the Ad Hoc Working Group assembled a scenario for human exploration, based largely on existing technology, which would have cost an estimated $400 to $500 billion. Regardless of the potential problems represented by this cost profile, the White House decided to go ahead with the announcement of SEI—probably assuming that cheaper alternatives could be found for the initiative.

On 20 July 1989, during his speech announcing SEI, President Bush directed the Space Council to assess technical approaches and budgetary resources required to carry out the initiative. After the address, however, the Council largely abdicated this authority to NASA. This took the form of the 90-Day Study. Given the outcome of the initial alternative generation process (e.g., selection of a technical approach that would require more than doubling the agency's budget), the decision to allow NASA to control the post-announcement alternative generation process was clearly a mistake. The decision not to include other actors within the space policy community ultimately presented serious problems. This outcome could have been avoided if a presidential decision directive with a detailed strategy for implementing SEI had been released concurrently with the president's speech. Instead, this guidance was not provided for nearly eight months. The inability of the Council staff to draft such a directive, given the short period of time available, presents a

---

[1] On 14 January 2004, President George W. Bush proposed a long-range plan for the American space program that included phasing out the aging Space Shuttle, redefining a partially constructed Space Station, and developing a crew exploration vehicle to return humans permanently to the Moon.

compelling explanation for why NASA was allowed to generate the 90-Day Study in virtual isolation with scant direction from the White House.

From the start, the Space Council should have more firmly controlled development of SEI. Given the administration's goals, after President Bush announced the initiative the Council should have commenced technical studies conducted by NASA and outside actors (e.g., government contractors, universities, think tanks, and national laboratories) based on detailed written guidance. This type of coordinated alternative generation process would have engendered the kind of "clean sheet" thinking the White House desired. Instead, the space agency followed a more expedient path and developed reference approaches based on past studies. Although the Space Council had verbally asked NASA to supply a variety of technical options and cost profiles, the top NASA leadership either misunderstood or ignored those requests. By the time the 90-Day Study was released, it was probably too late to regain control of the initiative. Congress already associated SEI with $500 billion budgetary requirements. Without any other studies initiated to provide real alternatives, the Council could not present a compelling argument that cheaper options existed. By the time the National Research Council and Synthesis Group were brought in to provide this perspective, the damage had already been done. Similarly, the presidential directives released in spring 1990 arrived too late to save the initiative.

Perhaps the most important lesson learned from the failure of SEI was that NASA needs competition for ideas from other space policy community actors. Since the post-Apollo deceleration, NASA has not proven itself capable of presenting White House policy makers with a robust suite of policy alternatives for large human spaceflight programs. From the earliest stages, the Space Council relied too much on the space agency to develop alternative approaches for SEI. There were a number of warning signs that should have led the council staff to bring other governmental and nongovernmental actors into the process. The most troubling of these was the initial $400 billion price tag introduced by the Ad Hoc Working Group. Given existing political and budgetary constraints, it is beyond explanation why the administration didn't seek out cheaper options before announcing the initiative. Instead, Vice President Quayle and Mark Albrecht were apparently satisfied with NASA's conclusion that the initiative was technically feasible. They chose not to focus on the staggering price tag at that point. Regardless, the political infeasibility of the required cost profile clearly demonstrated the need to include other actors in the subsequent alternative generation process. Additional warning signs that the alternative generation process had gone awry emerged even before NASA began assembling the 90-Day Study. The most obvious example was the conversation between Mark Albrecht and Aaron Cohen, when it became clear the two actors had fundamentally different definitions of alternatives. Throughout that period, the Space Council staff became increasingly alarmed by the lack of technical details being provided by NASA. Despite these growing concerns, however, the administration maintained its "wait and see" approach. By the time the 90-Day Study was released, it proved impossible to turn back the clock.

Responsibility for the ultimate demise of SEI should not all land at the White House doorstep. NASA missteps shared equally in its failure. Although it was directed to develop multiple options with different cost profiles, NASA presented only one expensive reference approach. To check the "alternatives" box, the agency simply provided slightly varied mission timelines and potential destinations. Even without clear written guidance from the administration, Admiral Truly and NASA's senior leadership should have recognized that existing budgetary constraints necessitated consideration of alternatives that could be implemented with modest resources. Instead, a plan was developed that never had any hope of gaining congressional support. An examination of the 90-Day Study reveals several factors that virtually guaranteed this outcome. First, the TSG selected an aggressive development sequence that called for emplacement, consolidation, and operations phases. While this may have been a logical strategy for maintaining permanent human presence beyond Earth orbit, it was not politically feasible. Second, making the ultimate objective of both Moon and Mars exploration the establishment of permanent outposts dramatically increased the expense of the initiative. Cheaper stand-alone missions should have been included. Third, making the Space Shuttle and Space Station programs central to the system architecture drove costs up dramatically. It also alienated potential supporters on Capitol Hill. Dick Malow said later he would have been "more positive if NASA had taken Space Station off the plate and focused on going to Mars. The budget envelope that would have opened up would have been sufficient to get the initiative going. I never felt space station was critical to going to Mars, never made any sense to me, if anything it may even have been a detractor."[2] Fourth, innovative in-space propulsion technologies were not given serious consideration. The basic chemical propulsion designs selected could have been supplemented with other options ranging from electric propulsion to solar sails to nuclear propulsion. Finally, additional strategic approaches should have been included. There was no shortage of architectural approaches available. In the end, the TSG's failure to consider a wide-variety of alternatives crippled SEI and exposed NASA to wide-ranging criticism. Although the Space Council attempted to find other options, the odds were already stacked against the initiative. In the end, the failure of the Space Council to coordinate a competition of ideas from the outset doomed the initiative.

---

[2] Malow interview.

Chapter 7: The Lessons of SEI

Even before he won the presidency, George Bush acknowledged that any new human spaceflight program would be significantly constrained by the federal budget. During the transition, the NASA Transition Team recommended the establishment of an agency priority-setting mechanism to take into account these concerns regarding the budget deficit. On his first day in office, President Bush told Congress his top agenda item was deficit reduction. Throughout its early months in office, the administration made clear that any decision regarding human spaceflight programs beyond Earth orbit would be made taking into account the limited resources available. In April 1989, when Vice President Quayle and Mark Albrecht met with Richard Darman and Bob Grady from OMB, it was agreed that any new initiative could not have a major budgetary impact. This body of evidence reveals that the Bush administration knew it would not gain congressional support for expensive exploration projects. It is equally clear, however, that NASA did not understand, or chose to ignore, this political reality. This is the primary reason why written guidance regarding the formulation of alternatives with favorable cost profiles was needed.

Drawing on the Apollo paradigm, NASA leaders believed President Bush's endorsement of a bold human spaceflight initiative was an opportunity to obtain a large funding increase. This was made clear when the AHWG developed a strategic architecture that would have required more than doubling the agency's budget. This was exactly what the administration did not want, yet Vice President Quayle and the Space Council perplexingly endorsed the review and proceeded with plans to announce SEI. Instead, given its desire to keep the space agency's budget in check, the administration should have taken one of two actions. First, it could have postponed the announcement of the initiative on the 20th anniversary of Apollo 11 and sought out less expensive options. Second, it could have announced SEI and immediately commenced a competition of ideas to determine what other alternatives were available. Rather than take control of the policy making process, the Space Council abdicated its authority to NASA. Admiral Truly, in turn, believed there was a right way and a wrong way of doing things, and the right way didn't include considering resource constraints. Preparing plans for SEI based on this fundamental principle proved to be an enormous political miscalculation.

Admiral Truly's approach to SEI was not without precedent. Something similar happened during post-Apollo planning. During that period, NASA advocated an aggressive human spaceflight program based on Vice President Agnew's support for Mars exploration. President Nixon and Congress were not won over. Rather than learn from this disappointment, the agency followed essentially the same course 20 years later. An examination of the record suggests the agency never seriously considered using President Bush's backing to gain support for modest budgetary increases that could fund a more limited human exploration program, which would not necessarily eliminate Mars missions. Thus, NASA missed a historic opportunity to right itself two decades after the post-Apollo deceleration. Instead, it proposed a highly expensive reference approach that would require doubling the agency budget because there was no inclination to cancel on-going programs. SEI's resultant demise badly damaged the American space program.

The failure to adequately consult Congress was one of the biggest mistakes made by the Bush administration before announcing SEI. Considering both houses were controlled by the opposition party, and given the existing budgetary crisis, this should have been a crucial part of the policy making process. Congress and congressional staffers are among the most critical actors in any policy community. Despite this fact, the administration did not involve legislators in SEI's agenda setting process. Instead, the Space Council and NASA simply 'informed' key members and staffers, instead of 'consulting' them regarding the initiative's substance or political feasibility. Furthermore, there was no attempt made to build a coalition of supporters for human exploration beyond Earth orbit. This explains why SEI never had any true champions on Capitol Hill, even among constituencies usually supportive of the space program. This procedural flaw was compounded when the Space Council and NASA continued to operate without seeking advice from Congress as alternatives were being generated. It is not surprising, therefore, that the 90-Day Study was pronounced 'dead-on-arrival' when it reached Capitol Hill. Without Congressional buy-in, it was impossible to garner support for SEI after the report was released. The Space Council made continual efforts to prove the program could be implemented with fewer resources. The space summit held in May 1990 was indicative of this struggle, but did not sway any Congressional supporters to come to the fore. By the time the Hubble flaw and Space Shuttle leaks were revealed, it was too late to save SEI. It is unclear whether broad-based support would have been forthcoming even if attempts had been made to build a coalition of congressional supporters, but the failure to make the effort clearly contributed to the ultimate failure of the initiative.

The philosophical disconnect between Admiral Truly and the Space Council was one of the most significant secondary causes of SEI's demise. After taking office, the Bush White House did not take early steps to find a new NASA Administrator. Although the administration knew James Fletcher was departing, there was no initial rush to find his replacement. Thus, the eventual decision to appoint Truly was made very rapidly. As a result, there was little time to make sure that his vision for the agency's future matched President Bush and Vice President Quayle's. Over the course of the subsequent three years, it became increasingly clear Truly's priorities were at odds with the Space Council's. In fact, Truly actively fought the Council's efforts to take control of space policy making within the federal government. In the

Chapter 7: The Lessons of SEI

end, this led to his firing and the hiring of Dan Goldin. SEI's outcome demonstrates how important it is for the president and NASA administrator to be on the same page when trying to gain approval for a major human spaceflight initiative.

In the absence of a true crisis environment, rapid decision making to meet arbitrary deadlines has not proven to be terribly successfully within the American space program. The determination to announce SEI on the 20th anniversary of the Apollo 11 landing was a perfect example. Although many within the space policy community recommended that NASA embark on an extended evaluation of future options, a decision to approve SEI was made in less than four months. As a result, President George Bush announced an initiative that had not been thoroughly examined with regard to its fiscal and political feasibility. In retrospect, it is unclear what necessitated this rushed process. After the speech, this perceived need for speed carried over to the 90-Day Study. An examination of the historical record does not reveal any clear rationale for conducting the study in three months, except to include funding for SEI in the fiscal year 1991 budget request. Regardless, the result was that the TSG did not have time to adequately evaluate a range of strategic architectures with different cost profiles. The NRC study team that reviewed the report criticized this quick turnaround as the proximate cause for the lack of real alternatives. This outcome virtually eliminated any chance for SEI's approval.

During the second half of the 20th century, there were a number of seminal moments in American space policy. These included the creation of NASA, President Kennedy's Moon decision, and the Space Shuttle and Space Station decisions. Due to its influence on the space program's future course, SEI rightfully belongs on this list. It is an anomaly in some respects because it was a failed initiative. Combined with the Hubble Space Telescope flaw and Space Shuttle fuel leaks, its demise led to significant changes at NASA. Perhaps the most important was the appointment of Dan Goldin, the most change-oriented administrator since James Webb.[3] The most important change he wrought was forcing NASA to face budgetary reality and focus on evolutionary advancement. This arguably wouldn't have happened absent the extraordinary budgetary requirements of NASA's SEI reference approach and the eventual downfall of the initiative.

The demise of SEI was a classic example of a defective decision-making process. The decision to conduct the agenda setting process in secret made it difficult to generate support within Congress or the space policy community. The Space Council's inability to provide high-level policy guidance, combined with NASA's failure to

---

[3] W. Henry Lambright, *Transforming Government: Dan Goldin and the Remaking of NASA* (Arlington, VA: PriceWaterhouseCoopers, 2001), p. 11.

independently consider critical fiscal constraints, derailed the initiative before it really got started. Finally, the failure of the Space Council to initiate a competition of ideas after President Bush's announcement speech removed any possibility of gaining congressional support after the devastating release of the 90-Day Study. It is far from obvious that the failure of SEI was predetermined given the existing budgetary crisis facing the nation in 1989. What is clear, however, is that its failure was ensured because options that may have been politically feasible were not considered during a deeply flawed policy process. While this had the benefit of forcing some level of change within NASA, it also badly damaged the agency's reputation as a world-class technical organization. To ensure the success of future efforts to send humans to Mars, current and future policy makers must learn the lessons of SEI. This alone is why its history is so fundamental to understanding what is required to gain support for large human spaceflight initiatives.

NASA History Series

Reference Works, NASA SP-4000:

Grimwood, James M. *Project Mercury: A Chronology*. NASA SP-4001, 1963.

Grimwood, James M., and C. Barton Hacker, with Peter J. Vorzimmer. *Project Gemini Technology and Operations: A Chronology*. NASA SP-4002, 1969.

Link, Mae Mills. *Space Medicine in Project Mercury*. NASA SP-4003, 1965.

Astronautics and Aeronautics, 1963: Chronology of Science, Technology, and Policy. NASA SP-4004, 1964.

Astronautics and Aeronautics, 1964: Chronology of Science, Technology, and Policy. NASA SP-4005, 1965.

Astronautics and Aeronautics, 1965: Chronology of Science, Technology, and Policy. NASA SP-4006, 1966.

Astronautics and Aeronautics, 1966: Chronology of Science, Technology, and Policy. NASA SP-4007, 1967.

Astronautics and Aeronautics, 1967: Chronology of Science, Technology, and Policy. NASA SP-4008, 1968.

Ertel, Ivan D., and Mary Louise Morse. *The Apollo Spacecraft: A Chronology, Volume I, Through November 7, 1962*. NASA SP-4009, 1969.

Morse, Mary Louise, and Jean Kernahan Bays. *The Apollo Spacecraft: A Chronology, Volume II, November 8, 1962–September 30, 1964*. NASA SP-4009, 1973.

Brooks, Courtney G., and Ivan D. Ertel. *The Apollo Spacecraft: A Chronology, Volume III, October 1, 1964–January 20, 1966*. NASA SP-4009, 1973.

Ertel, Ivan D., and Roland W. Newkirk, with Courtney G. Brooks. *The Apollo Spacecraft: A Chronology, Volume IV, January 21, 1966–July 13, 1974*. NASA SP-4009, 1978.

Astronautics and Aeronautics, 1968: Chronology of Science, Technology, and Policy. NASA SP-4010, 1969.

Newkirk, Roland W., and Ivan D. Ertel, with Courtney G. Brooks. *Skylab: A Chronology.* NASA SP-4011, 1977.

Van Nimmen, Jane, and Leonard C. Bruno, with Robert L. Rosholt. *NASA Historical Data Book, Volume I: NASA Resources, 1958–1968.* NASA SP-4012, 1976, rep. ed. 1988.

Ezell, Linda Neuman. *NASA Historical Data Book, Volume II: Programs and Projects, 1958–1968.* NASA SP-4012, 1988.

Ezell, Linda Neuman. *NASA Historical Data Book, Volume III: Programs and Projects, 1969–1978.* NASA SP-4012, 1988.

Gawdiak, Ihor Y., with Helen Fedor, compilers. *NASA Historical Data Book, Volume IV: NASA Resources, 1969–1978.* NASA SP-4012, 1994.

Rumerman, Judy A., compiler. *NASA Historical Data Book, 1979–1988: Volume V, NASA Launch Systems, Space Transportation, Human Spaceflight, and Space Science.* NASA SP-4012, 1999.

Rumerman, Judy A., compiler. *NASA Historical Data Book, Volume VI: NASA Space Applications, Aeronautics and Space Research and Technology, Tracking and Data Acquisition/Space Operations, Commercial Programs, and Resources, 1979–1988.* NASA SP-2000-4012, 2000.

Astronautics and Aeronautics, 1969: Chronology of Science, Technology, and Policy. NASA SP-4014, 1970.

Astronautics and Aeronautics, 1970: Chronology of Science, Technology, and Policy. NASA SP-4015, 1972.

Astronautics and Aeronautics, 1971: Chronology of Science, Technology, and Policy. NASA SP-4016, 1972.

Astronautics and Aeronautics, 1972: Chronology of Science, Technology, and Policy. NASA SP-4017, 1974.

Astronautics and Aeronautics, 1973: Chronology of Science, Technology, and Policy. NASA SP-4018, 1975.

Astronautics and Aeronautics, 1974: Chronology of Science, Technology, and Policy.

NASA SP-4019, 1977.

Astronautics and Aeronautics, 1975: Chronology of Science, Technology, and Policy. NASA SP-4020, 1979.

Astronautics and Aeronautics, 1976: Chronology of Science, Technology, and Policy. NASA SP-4021, 1984.

Astronautics and Aeronautics, 1977: Chronology of Science, Technology, and Policy. NASA SP-4022, 1986.

Astronautics and Aeronautics, 1978: Chronology of Science, Technology, and Policy. NASA SP-4023, 1986.

Astronautics and Aeronautics, 1979–1984: Chronology of Science, Technology, and Policy. NASA SP-4024, 1988.

Astronautics and Aeronautics, 1985: Chronology of Science, Technology, and Policy. NASA SP-4025, 1990.

Noordung, Hermann. *The Problem of Space Travel: The Rocket Motor. Edited by Ernst Stuhlinger and J. D. Hunley, with Jennifer Garland.* NASA SP-4026, 1995.

Astronautics and Aeronautics, 1986–1990: A Chronology. NASA SP-4027, 1997.

Astronautics and Aeronautics, 1990–1995: A Chronology. NASA SP-2000-4028, 2000.

Management Histories, NASA SP-4100:

Rosholt, Robert L. *An Administrative History of NASA, 1958–1963.* NASA SP-4101, 1966.

Levine, Arnold S. *Managing NASA in the Apollo Era.* NASA SP-4102, 1982.

Roland, Alex. *Model Research: The National Advisory Committee for Aeronautics, 1915–1958.* NASA SP-4103, 1985.

Fries, Sylvia D. *NASA Engineers and the Age of Apollo.* NASA SP-4104, 1992.

Glennan, T. Keith. *The Birth of NASA: The Diary of T. Keith Glennan. J. D. Hunley, editor.* NASA SP-4105, 1993.

Seamans, Robert C., Jr. *Aiming at Targets: The Autobiography of Robert C. Seamans, Jr.* NASA SP-4106, 1996.

Garber, Stephen J., editor. *Looking Backward, Looking Forward: Forty Years of U.S. Human Spaceflight Symposium.* NASA SP-2002-4107, 2002.

Mallick, Donald L. with Peter W. Merlin. *The Smell of Kerosene: A Test Pilot's Odyssey.* NASA SP-4108, 2003.

Iliff, Kenneth W. and Curtis L. Peebles. *From Runway to Orbit: Reflections of a NASA Engineer.* NASA SP-2004-4109, 2004.

Chertok, Boris. *Rockets and People, Volume 1.* NASA SP-2005-4110, 2005.

Laufer, Alexander, Todd Post, and Edward Hoffman. *Shared Voyage: Learning and Unlearning from Remarkable Projects.* NASA SP-2005-4111, 2005.

Dawson, Virginia P. and Mark D. Bowles. *Realizing the Dream of Flight: Biographical Essays in Honor of the Centennial of Flight, 1903-2003.* NASA SP-2005-4112, 2005.

Project Histories, NASA SP-4200:

Swenson, Loyd S., Jr., James M. Grimwood, and Charles C. Alexander. *This New Ocean: A History of Project Mercury.* NASA SP-4201, 1966; rep. ed. 1998.

Green, Constance McLaughlin, and Milton Lomask. *Vanguard: A History.* NASA SP-4202, 1970; rep. ed. Smithsonian Institution Press, 1971.

Hacker, Barton C., and James M. Grimwood. *On the Shoulders of Titans: A History of Project Gemini.* NASA SP-4203, 1977.

Benson, Charles D., and William Barnaby Faherty. *Moonport: A History of Apollo Launch Facilities and Operations.* NASA SP-4204, 1978.

Brooks, Courtney G., James M. Grimwood, and Loyd S. Swenson, Jr. *Chariots for Apollo: A History of Manned Lunar Spacecraft.* NASA SP-4205, 1979.

Bilstein, Roger E. *Stages to Saturn: A Technological History of the Apollo/Saturn Launch Vehicles.* NASA SP-4206, 1980, rep. ed. 1997.

SP-4207 not published.

Compton, W. David, and Charles D. Benson. *Living and Working in Space: A History of Skylab.* NASA SP-4208, 1983.

Ezell, Edward Clinton, and Linda Neuman Ezell. *The Partnership: A History of the Apollo-Soyuz Test Project.* NASA SP-4209, 1978.

Hall, R. Cargill. *Lunar Impact: A History of Project Ranger.* NASA SP-4210, 1977.

Newell, Homer E. B*eyond the Atmosphere: Early Years of Space Science.* NASA SP-4211, 1980.

Ezell, Edward Clinton, and Linda Neuman Ezell. *On Mars: Exploration of the Red Planet, 1958–1978.* NASA SP-4212, 1984.

Pitts, John A. *The Human Factor: Biomedicine in the Manned Space Program to 1980.* NASA SP-4213, 1985.

Compton, W. David. *Where No Man Has Gone Before: A History of Apollo Lunar Exploration Missions.* NASA SP-4214, 1989.

Naugle, John E. *First Among Equals: The Selection of NASA Space Science Experiments.* NASA SP-4215, 1991.

Wallace, Lane E. *Airborne Trailblazer: Two Decades with NASA Langley's Boeing 737 Flying Laboratory.* NASA SP-4216, 1994.

Butrica, Andrew J., editor. *Beyond the Ionosphere: Fifty Years of Satellite Communication.* NASA SP-4217, 1997.

Butrica, Andrew J. *To See the Unseen: A History of Planetary Radar Astronomy.* NASA SP-4218, 1996.

Mack, Pamela E., editor. *From Engineering Science to Big Science: The NACA and NASA Collier Trophy Research Project Winners.* NASA SP-4219, 1998.

Reed, R. Dale, with Darlene Lister. *Wingless Flight: The Lifting Body Story.* NASA SP-4220, 1997.

Heppenheimer, T. A. *The Space Shuttle Decision: NASA's Search for a Reusable Space Vehicle.* NASA SP-4221, 1999.

Hunley, J. D., editor. *Toward Mach 2: The Douglas D-558 Program.* NASA SP-4222, 1999.

Swanson, Glen E., editor. *"Before this Decade Is Out . . .": Personal Reflections on the Apollo Program*. NASA SP-4223, 1999.

Tomayko, James E. *Computers Take Flight: A History of NASA's Pioneering Digital Fly-by-Wire Project*. NASA SP-2000-4224, 2000.

Morgan, Clay. *Shuttle-Mir: The U.S. and Russia Share History's Highest Stage*. NASA SP-2001-4225, 2001.

Leary, William M. *"We Freeze to Please": A History of NASA's Icing Research Tunnel and the Quest for Flight Safety*. NASA SP-2002-4226, 2002.

Mudgway, Douglas J. *Uplink-Downlink: A History of the Deep Space Network 1957–1997*. NASA SP-2001-4227, 2001.

Dawson, Virginia P. and Mark D. Bowles. *Taming Liquid Hydrogen: The Centaur Upper Stage Rocket, 1958-2002*. NASA SP-2004-4230, 2004.

Meltzer, Michael. *Mission to Jupiter: A History of the Galileo Project*. NASA SP-2007-4231.

Center Histories, NASA SP-4300:

Rosenthal, Alfred. *Venture into Space: Early Years of Goddard Space Flight Center*. NASA SP-4301, 1985.

Hartman, Edwin P. *Adventures in Research: A History of Ames Research Center, 1940–1965*. NASA SP-4302, 1970.

Hallion, Richard P. *On the Frontier: Flight Research at Dryden, 1946–1981*. NASA SP-4303, 1984.

Muenger, Elizabeth A. *Searching the Horizon: A History of Ames Research Center, 1940–1976*. NASA SP-4304, 1985.

Hansen, James R. *Engineer in Charge: A History of the Langley Aeronautical Laboratory, 1917–1958*. NASA SP-4305, 1987.

Dawson, Virginia P. *Engines and Innovation: Lewis Laboratory and American Propulsion Technology*. NASA SP-4306, 1991.

Dethloff, Henry C. *"Suddenly Tomorrow Came . . .": A History of the Johnson Space Center*. NASA SP-4307, 1993.

Hansen, James R. *Spaceflight Revolution: NASA Langley Research Center from Sputnik to Apollo*. NASA SP-4308, 1995.

Wallace, Lane E. *Flights of Discovery: 50 Years at the NASA Dryden Flight Research Center*. NASA SP-4309, 1996.

Herring, Mack R. *Way Station to Space: A History of the John C. Stennis Space Center*. NASA SP-4310, 1997.

Wallace, Harold D., Jr. *Wallops Station and the Creation of the American Space Program*. NASA SP-4311, 1997.

Wallace, Lane E. *Dreams, Hopes, Realities: NASA's Goddard Space Flight Center, The First Forty Years*. NASA SP-4312, 1999.

Dunar, Andrew J., and Stephen P. Waring. *Power to Explore: A History of the Marshall Space Flight Center*. NASA SP-4313, 1999.

Bugos, Glenn E. *Atmosphere of Freedom: Sixty Years at the NASA Ames Research Center*. NASA SP-2000-4314, 2000.

Schultz, James. *Crafting Flight: Aircraft Pioneers and the Contributions of the Men and Women of NASA Langley Research Center*. NASA SP-2003-4316, 2003.

General Histories, NASA SP-4400:

Corliss, William R. *NASA Sounding Rockets, 1958–1968: A Historical Summary*. NASA SP-4401, 1971.

Wells, Helen T., Susan H. Whiteley, and Carrie Karegeannes. *Origins of NASA Names*. NASA SP-4402, 1976.

Anderson, Frank W., Jr. *Orders of Magnitude: A History of NACA and NASA, 1915–1980*. NASA SP-4403, 1981.

Sloop, John L. *Liquid Hydrogen as a Propulsion Fuel, 1945–1959*. NASA SP-4404, 1978.

Roland, Alex. *A Spacefaring People: Perspectives on Early Spaceflight*. NASA SP-4405, 1985.

Bilstein, Roger E. *Orders of Magnitude: A History of the NACA and NASA, 1915–

1990. NASA SP-4406, 1989.

Logsdon, John M., editor, with Linda J. Lear, Jannelle Warren-Findley, Ray A. Williamson, and Dwayne A. Day. *Exploring the Unknown: Selected Documents in the History of the U.S. Civil Space Program, Volume I, Organizing for Exploration*. NASA SP-4407, 1995.

Logsdon, John M., editor, with Dwayne A. Day and Roger D. Launius. *Exploring the Unknown: Selected Documents in the History of the U.S. Civil Space Program, Volume II, Relations with Other Organizations*. NASA SP-4407, 1996.

Logsdon, John M., editor, with Roger D. Launius, David H. Onkst, and Stephen J. Garber. *Exploring the Unknown: Selected Documents in the History of the U.S. Civil Space Program, Volume III, Using Space*. NASA SP-4407, 1998.

Logsdon, John M., general editor, with Ray A. Williamson, Roger D. Launius, Russell J. Acker, Stephen J. Garber, and Jonathan L. Friedman. *Exploring the Unknown: Selected Documents in the History of the U.S. Civil Space Program, Volume IV, Accessing Space*. NASA SP-4407, 1999.

Logsdon, John M., general editor, with Amy Paige Snyder, Roger D. Launius, Stephen J. Garber, and Regan Anne Newport. *Exploring the Unknown: Selected Documents in the History of the U.S. Civil Space Program, Volume V, Exploring the Cosmos*. NASA SP-2001-4407, 2001.

Siddiqi, Asif A. *Challenge to Apollo: The Soviet Union and the Space Race, 1945–1974*. NASA SP-2000-4408, 2000.

Hansen, James R., editor. *The Wind and Beyond: Journey into the History of Aerodynamics in America, Volume 1, The Ascent of the Airplane*. NASA SP-2003-4409, 2003.

Monographs in Aerospace History, NASA SP-4500:

Launius, Roger D. and Aaron K. Gillette, compilers, *Toward a History of the Space Shuttle: An Annotated Bibliography*. Monograph in Aerospace History, No. 1, 1992.

Launius, Roger D., and J. D. Hunley, compilers, *An Annotated Bibliography of the Apollo Program*. Monograph in Aerospace History, No. 2, 1994.

NASA History Series

Launius, Roger D. Apollo: A Retrospective Analysis. Monograph in Aerospace History, No. 3, 1994.

Hansen, James R. *Enchanted Rendezvous: John C. Houbolt and the Genesis of the Lunar-Orbit Rendezvous Concept.* Monograph in Aerospace History, No. 4, 1995.

Gorn, Michael H. *Hugh L. Dryden's Career in Aviation and Space.* Monograph in Aerospace History, No. 5, 1996.

Powers, Sheryll Goecke. *Women in Flight Research at NASA Dryden Flight Research Center, from 1946 to 1995.* Monograph in Aerospace History, No. 6, 1997.

Portree, David S. F. and Robert C. Trevino. *Walking to Olympus: An EVA Chronology.* Monograph in Aerospace History, No. 7, 1997.

Logsdon, John M., moderator. *Legislative Origins of the National Aeronautics and Space Act of 1958: Proceedings of an Oral History Workshop.* Monograph in Aerospace History, No. 8, 1998.

Rumerman, Judy A., compiler, *U.S. Human Spaceflight, A Record of Achievement 1961–1998.* Monograph in Aerospace History, No. 9, 1998.

Portree, David S. F. *NASA's Origins and the Dawn of the Space Age.* Monograph in Aerospace History, No. 10, 1998.

Logsdon, John M. *Together in Orbit: The Origins of International Cooperation in the Space Station.* Monograph in Aerospace History, No. 11, 1998.

Phillips, W. Hewitt. *Journey in Aeronautical Research: A Career at NASA Langley Research Center.* Monograph in Aerospace History, No. 12, 1998.

Braslow, Albert L. *A History of Suction-Type Laminar-Flow Control with Emphasis on Flight Research.* Monograph in Aerospace History, No. 13, 1999.

Logsdon, John M., moderator. *Managing the Moon Program: Lessons Learned From Apollo.* Monograph in Aerospace History, No. 14, 1999.

Perminov, V. G. *The Difficult Road to Mars: A Brief History of Mars Exploration in the Soviet Union.* Monograph in Aerospace History, No. 15, 1999.

Tucker, Tom. *Touchdown: The Development of Propulsion Controlled Aircraft at NASA Dryden.* Monograph in Aerospace History, No. 16, 1999.

Maisel, Martin D., Demo J. Giulianetti, and Daniel C. Dugan. *The History of the XV-15 Tilt Rotor Research Aircraft: From Concept to Flight.* NASA SP-2000-4517, 2000.

Jenkins, Dennis R. *Hypersonics Before the Shuttle: A Concise History of the X-15 Research Airplane.* NASA SP-2000-4518, 2000.

Chambers, Joseph R. *Partners in Freedom: Contributions of the Langley Research Center to U.S. Military Aircraft in the 1990s.* NASA SP-2000-4519, 2000.

Waltman, Gene L. *Black Magic and Gremlins: Analog Flight Simulations at NASA's Flight Research Center.* NASA SP-2000-4520, 2000.

Portree, David S. F. *Humans to Mars: Fifty Years of Mission Planning, 1950–2000.* NASA SP-2001-4521, 2001.

Thompson, Milton O., with J. D. Hunley. *Flight Research: Problems Encountered and What They Should Teach Us.* NASA SP-2000-4522, 2000.

Tucker, Tom. *The Eclipse Project.* NASA SP-2000-4523, 2000.

Siddiqi, Asif A. *Deep Space Chronicle: A Chronology of Deep Space and Planetary Probes, 1958–2000.* NASA SP-2002-4524, 2002.

Merlin, Peter W. *Mach 3+: NASA/USAF YF-12 Flight Research, 1969–1979.* NASA SP-2001-4525, 2001.

Anderson, Seth B. *Memoirs of an Aeronautical Engineer—Flight Tests at Ames Research Center: 1940–1970.* NASA SP-2002-4526, 2002.

Renstrom, Arthur G. *Wilbur and Orville Wright: A Bibliography Commemorating the One-Hundredth Anniversary of the First Powered Flight on December 17, 1903.* NASA SP-2002-4527, 2002.

No monograph 28.

Chambers, Joseph R. *Concept to Reality: Contributions of the NASA Langley Research Center to U.S. Civil Aircraft of the 1990s.* SP-2003-4529, 2003.

Peebles, Curtis, editor. *The Spoken Word: Recollections of Dryden History, The Early Years.* SP-2003-4530, 2003.

Jenkins, Dennis R., Tony Landis, and Jay Miller. *American X-Vehicles: An Inventory-X-1 to X-50.* SP-2003-4531, 2003.

Renstrom, Arthur G. *Wilbur and Orville Wright: A Chronology Commemorating the One-Hundredth Anniversary of the First Powered Flight on December 17, 1903.* NASA SP-2003-4532, 2002.

NASA History Series

Bowles, Mark D. and Robert S. Arrighi. *NASA's Nuclear Frontier: The Plum Brook Research Reactor.* SP-2004-4533, 2003.

Matranga, Gene J. and C. Wayne Ottinger, Calvin R. Jarvis with D. Christian Gelzer. *Unconventional, Contrary, and Ugly: The Lunar Landing Research Vehicle.* NASA SP-2006-4535.

McCurdy, Howard E. *Low Cost Innovation in Spaceflight: The History of the Near Earth Asteroid Rendezvous (NEAR) Mission.* NASA SP-2005-4536, 2005.

Seamans, Robert C. Jr. *Project Apollo: The Tough Decisions.* NASA SP-2005-4537, 2005.

Lambright, W. Henry. *NASA and the Environment: The Case of Ozone Depletion.* NASA SP-2005-4538, 2005.

Chambers, Joseph R. *Innovation in Flight: Research of the NASA Langley Research Center on Revolutionary Advanced Concepts for Aeronautics.* NASA SP-2005-4539, 2005.

Phillips, W. Hewitt. *Journey Into Space Research: Continuation of a Career at NASA Langley Research Center.* NASA SP-2005-4540, 2005.

Index

90-Day Study, 79-95, 96, 97, 99, 100, 102, 104-105, 108, 111, 115, 117, 129, 141, 143, 152, 155, 159, 161-162, 164, 165, 166

A
Aaron, John, 32, 59, 60
Adams, Robert, 69
Ad Hoc Working Group (AHWG), 59-62, 79, 83, 86, 160, 161, 163
Advanced Launch System (ALS), 88, 122
Aerospace plane, 29, 46, 111
Agenda setting, 14, 68, 86, 137, 138-143, 155, 157, 159, 164, 165
Agnew, Spiro, 21-22, 163
Albrecht, Mark, 2, 4, 48-49, 51, 52, 54, 57-58, 62-63, 65-66, 68-69, 79-82, 92-94, 95, 96, 99, 100, 105, 109, 114, 123, 127, 130-133, 139, 160-161, 163
Aldridge, Pete, 99
Aldrin, Buzz, 50, 69, 70, 120, 152
Alternative generation, 19, 59, 84, 94, 101, 138-143, 156, 159, 160-161
Anderson, Clinton, 23
Apollo, see Project Apollo
Apollo-Soyuz Test Project (ASTP), 53, 116, 145
Aristarchus of Samos, 6-7
Armstrong, Neil, 28, 50, 69, 70, 152
Astronauts, 10, 17, 19, 27, 28, 30, 31, 32, 40, 50, 53, 54, 61, 64, 72, 74, 84, 99, 116, 120, 149
Augustine, Norman, 78, 124, 126, 128

B
Baker, David, 117
Barnard, Edward, 8
Baumgartner, Frank, 3, 4-5, 137, 143-149, 148, 153, 156
Beggs, Jim, 28, 130
Bentsen, Lloyd, 39
Blue Ribbon Discussion Group, 99-102
Boston, Penelope, 25-26

179

Bradbury, Ray, 10
Brahe, Tycho, 6-7
Branscome, Darrell, 59, 60, 64
Bromley, Allan, 99
Burrough, Bryan, 81, 132
Burroughs, Edgar Rice, 9
Bush, George, 38-39, 42, 47, 68, 103, 109, 112, 141, 153, 163
   and 1988 presidential election, 40-44
   and 1992 presidential election, 132, 133
   and announcement of SEI, 1, 2, 15, 49-50, 53, 68-72, 74-5, 79, 138, 164, 165
   and *Challenger* families, 40
   and Vice President Dan Quayle, 52, 53, 56, 68, 69, 70, 72, 79, 107, 113, 141, 150, 164
   and international space cooperation, 109-110
   and Mark Albrecht, 48, 131
   and NASA Administrator James Fletcher, 40-41, 42, 44, 49, 164
   and NASA Administrator Richard Truly, 54, 130-31, 164
   and National Space Council, 56, 160, 162-163
   Reaction to 90-Day Study, 93
   Reagan-Bush transition, 44-46
   SEI Presidential decision directives, 108-110
   Speech at Marshall Space Flight Center, 118-119
   Speech at Texas A&I University, 114-115
   as supporter of Moon-Mars exploration, 42-44, 46, 47, 50, 92, 155, 159
   as supporter of space program, 38, 40, 42-44, 47, 107, 112, 114-115, 118-119, 138, 141, 150, 155, 159, 163
   as supporter of space station, 42-44, 47
   and White House space summit, 112-114
Bush, Prescott, 38

C
Campbell, Thomas, 111
Carter, Jimmy, 39
Case for Mars conferences, see Mars exploration
Cassini, Giovanni, 7
Center for Strategic and International Studies, 51
Cernan, Eugene, 120, 123
Clinton, Bill, 133, 134-135, 146
Cohen, Aaron, 2, 59, 79-81, 93-94, 161
Cohen, Michael, 3
Collier's magazine, 17-18

Collins, Michael, 50, 64, 65, 69, 70, 99
Columbus, Christopher, 72, 119
Congress, see United States Congress
Cooper, Henry, 48
Copernicus, Nicolaus, 6-7
Covault, Craig, 69
Craig, Mark, 59, 60, 62, 64, 65, 69, 82, 83
Culbertson, Phillip, 58

D
Darman, Richard, 52-53, 93, 96, 99, 100, 126, 139, 163
Darwin, Charles, 59, 60
David, Leonard, 26
Deimos, 7, 27
Demisch, Wolfgang, 78
Department of Energy, 109, 116
Department of Defense, 108-109, 116
Department of Transportation, 109
Disney, Walt, 18
Dole, Bob, 120
DuBridge, Lee, 21
Dukakis, Mike, 42, 44
Duke, Mike, 59, 60

E
Earth Observation System, 125, 132
Ehricke, Krafft, 20
EMPIRE study, 20
Energia launch vehicle, 41
Erlichman, John, 23

F
Faget, Maxime, 20
Fisk, Lennard, 120
Fitzwater, Marlin, 48, 73
Flammarion, Camille, 8
Fletcher, James, 30, 32-33, 40-42, 44, 45, 49-50, 51, 53, 130, 164
Ford Aeronutronic, 20
Ford, Gerald, 39
Franks, Anthony, 70
Friedman, Louis, 64

G
Galilei, Galileo, 7
Garbage Can Theory, 3-4
Garn, Jake, 47, 66, 112, 120
General Dynamics, 20
Geocentric model, 6-7
Gibson, Roy, 128
Gingrich, Newt, 120
Goldin, Dan, 132-134, 165
Gorbachev, Mikhail, 109-110
Gore, Al, 77, 115, 122, 131, 134
Grady, Bob, 52, 163
Gramm, Phil, 120
Green, Bill, 78
Greene, John Robert, 38-39

H
Hall, Asaph, 7
Heinlein, Robert, 9-10
Heliocentric model, 6-7
Herschel, William, 7
Hoff, Joan, 21-25
Hollings, Ernest, 67
Hubble Space Telescope, 2, 113, 118, 120-121, 165
Hunt, Guy, 118
Huygens, Christian, 7

J
Jastrow, Robert, 120
Johnson, Lyndon, 24, 39, 53, 148, 154
Jones, Bryan, 3, 4-5, 137, 143-149, 148, 153, 156

K
Kennedy, John, 19, 24, 34, 42, 53, 69, 78, 81, 148, 151, 153, 154, 165
Kepler, Johannes, 6-7
Keyworth, George, 28
Kingdon, John, 3-4, 137, 138-143, 157
Kirkpatrick, Jean, 28
Kohashi, Stephen, 49, 55, 66, 97, 125
Kristol, Bill, 123

L

Lawrence Livermore National Laboratory, 95-96, 99
Lenoir, Bill, 60, 62
Lewis, Tom, 119
Ley, Willy, 17
Lockheed, 20
Logsdon, John, 25, 35, 129, 131
Lovelace, Alan, 110
Lowell, Percival, 8-9, 10, 11
Lunar Observer mission, 86, 98, 122

M

Malow, Richard, 55, 66-67, 97, 109, 115, 121, 125, 128, 162
March, James, 3
Mariner program, 11-13, 16, 20
Mark, Hans, 123
Mars exploration, 15, 37, 42, 43, 45, 46, 50, 68-69, 71, 75, 93, 105, 114, 118-119, 125, 127, 132, 135, 143, 145, 146, 150, 151, 152, 154, 155, 159, 163
  and the Ad Hoc Working Group, 57-64
  Astronomical observations, 6-9
  and Augustine Commission Report, 126-127
  Canals on, 7-9
  Case for Mars conferences, 25-27, 117
  and competition with Soviet Union, 41, 108, 110
  Early mission planning, 16-21
  and Lawrence Livermore National Laboratory, 95-6
  Lunar and Mars Exploration Office, 59, 60
  Mariner missions, 11
  Mars Direct concept, 117
  Mars Global Network, 86
  Mars Observer, 98, 111
  Mars Polar Orbiter, 27
  Mars Rover, 31, 86
  Mars Sample Return, 27, 31, 86, 100
  Mars Site Reconnaissance Orbiter, 86
  and NASA Office of Exploration, 32-35
  in National Commission on Space, 27-30
  Office of Exploration, 32-35, 57, 59, 60, 102, 142
  in popular culture, 9-10
  Post-Apollo planning, 21-25

Mars Wars

    Rationales for exploration, 5-7, 72
    in Ride Report, 30-32
    as seen by early civilizations, 5-6
    and space policy community, 65-67
    and Synthesis Group Report, 128-129
    and the Technical Study Group, 82-92
    Viking missions, 12-13
Mars Society, 27
Mars Underground, 25-26
Martin, Frank, 57-58, 59, 62-64, 65, 67, 79, 81
Matsunaga, Spark, 112
Mayo, Robert, 23
McCurdy, Howard, 18, 19, 35
McCurry, Mike, 135
McKay, Chris, 25-26
Meyer, Tom, 25-26
Michener, James, 75
Mikulski, Barbara, 67, 112
Miller, George, 23
Mineta, Norman, 55
Mir space station, 41, 81, 132
Mission to Planet Earth, 31, 42, 62, 103, 113, 122, 126-127
Mitchell, Brad, 45-46
Moon exploration, 1, 2, 17, 18, 19, 29, 30, 31, 32, 33, 35, 41, 42, 43, 45, 46, 52, 57, 58, 59, 60, 62, 63-64, 65, 66, 68, 69, 71-73, 74, 75, 77, 85-86, 89, 90-91, 93, 95, 97, 98, 100, 101, 102, 103, 105, 108, 111, 112, 113, 119, 123, 124, 129, 134, 144, 145, 148, 150, 151, 159, 162
Murray, Bruce, 64-65, 123

N

NASA, 2, 11, 14, 19, 21, 22, 37, 38, 41, 42, 50, 134
    90-Day Study, 82-92
    and Ad Hoc Working Group, 57-64, 68-69
    and Augustine Commission, 123-124, 127-128
    and Blue Ribbon Discussion Group, 99-100
    Budget, 23, 30, 42, 46, 47, 62, 63, 64, 66, 67, 73, 78, 80, 92, 94-95, 97-99, 103-104, 110-111, 121-122, 124-127, 152-3, 155, 159
    and Center for Strategic and International Studies, 51
    and *Challenger* accident, 38, 40, 44
    History Division, 146
    and Hubble Space Telescope flaw, 120-121

Jet Propulsion Laboratory, 60, 91, 127
Johnson Space Center, 2, 20, 59, 60, 63, 81-84, 91, 97, 99, 104, 151
and lessons learned from SEI, 159-166
Lewis Research Center, 19, 132
Manned Spacecraft Center, 20
Marshall Space Flight Center, 19, 20, 41, 91, 118-119
and Mitchell Report, 45-46
and National Commission on Space, 27-30
and National Research Council, 45, 104-105
and National Space Council, 2, 48, 49, 52, 79-84, 92-105, 108-110, 114, 131
Office of Exploration, 32-35
and Post-Apollo planning, 21-25
and public opinion, 146-147
and Ride Report, 30-32
and space policy community, 64-67, 138-143
and Synthesis Group, 128-130
Technical Study Group, see 90-Day Study
NASA Transition Office Team, 45-46
National Air and Space Museum, 1, 50, 69
National Academy of Sciences, 96, 100, 104-105, 108, 110, 111, 116, 161, 165
National Commission on Space, 27-30, 34, 37, 59, 65, 145
National Research Council, see National Academy of Sciences
National Science and Technology Council, 134
National Security Council, 34
National Space Council, 1, 2, 46, 47, 48, 49, 50, 52, 54, 56-57, 58, 60, 68, 72, 74, 77, 79-85, 92-97, 99-101, 103-104, 108-110, 111, 113, 114, 115, 116, 123, 126-128, 130, 131, 132, 133, 134, 139, 141, 142, 150, 152, 159, 160-166
Nelson, Bill, 128
Nixon, Richard, 15, 21-24, 25, 39, 146, 148, 152, 154, 159, 163

O
Office of Science and Technology Policy, 139
Office of Technology Assessment, 128
O'Handley, Douglas, 60, 62, 65, 79, 82, 83, 84, 93, 96
Olsen, Johan, 3

P
Paine, Thomas, 21-22, 27-28, 32, 64-65, 99, 123, 128, 130
Panetta, Leon, 78, 134
Pfiffner, James, 44
Phobos, 7, 27, 33

Pike, John, 48, 104
Policy entrepreneur, 4, 22, 49, 52, 57, 138-140, 159
Policy Streams Model, 3-4, 137, 138-143, 155-156
Policy window, 3, 4, 44, 146, 160
Portree, David, 151
Post-Apollo Planning, see Project Apollo
Project Apollo, 1, 20, 22, 26, 31, 37, 64, 78, 79, 81, 115, 119, 143, 145, 149, 152-3, 154, 155
    Apollo 11 anniversary, 1, 2, 50, 51, 57, 69-75, 163, 165
    Post-Apollo Planning, 2, 5, 14, 15, 20, 21-5, 84, 145, 145, 152, 155, 159, 163
Proxmire, William, 66
Ptolemy, Claudius, 6-7
Pulliam, Eugene, 52
Punctuated Equilibrium Model, 3, 4-5, 137, 143-149, 155, 157

Q
Quayle, Dan, 2, 52
    and announcement of SEI, 69, 70, 72, 79
    as Chair, National Space Council, 46, 47, 49, 52, 56, 60, 62-63, 67, 68-69, 80, 94-96, 99, 100, 102, 108, 113, 116-117, 120, 122-123, 125, 126-127, 129, 150, 154, 161, 163, 164
    and Dan Goldin, 132-133
    and development of SEI, 51-53
    and George Bush, 52, 53, 56, 68, 70, 72, 79, 107, 113, 164
    and Mark Albrecht, 4, 49, 51, 79, 93, 160-161, 163
    and reaction to 90-Day Study, 93
    and Richard Truly, 53-54, 57, 67, 117, 130-131
    and space policy community, 138-141

R
RAND Corporation, 48, 116
Reagan, Ronald, 16, 28, 34-35, 38, 39, 41, 44, 46, 66, 110, 150, 152, 154
Rice, Donald, 96
Ride, Sally, 30-32
Russia, see Soviet Union
Ryan, Cornelius, 17

S
Sagan, Carl, 99, 109
Sasser, James, 77
Sawyer, Kathy, 95

Index

Schiaparelli, Virginio, 8, 10
Schmitt, Harrison, 99
Scobee, June, 40
Scowcroft, Brent, 50, 51
Seamans, Robert, 21, 23
Senior Interagency Group-Space, 34
Shuttle-C, 61
Skinner, Samuel, 130
Skylab, 25, 53, 145
Smith, Marcia, 35
Smithsonian Institution, see National Air and Space Museum
Softening up process, 16, 17, 19, 27, 145, 151
Soviet Union, 11, 12, 24, 30, 37, 41, 43, 51, 107, 109-110, 134, 152
Space policy community, 3, 27, 37, 44, 64, 84, 96, 117, 132, 138-143, 157, 160-161, 165
Space shuttle, 2, 24, 25, 26, 27, 28, 29, 34, 37, 38, 40, 41, 42, 47, 53, 54, 59, 61, 84, 105, 113, 121, 122, 125, 126, 127, 128, 131, 143, 144, 145, 146, 150, 151, 154, 162, 164, 165
 and *Challenger* accident, 28, 30, 34, 38, 40, 42, 44, 54, 72, 150-151, 154
 Space station, 1, 2, 17, 20, 22, 24, 28, 31, 34, 37, 41, 43, 45, 46, 47, 50, 52, 53, 61, 62, 71, 75, 83-89, 97, 98, 103, 111, 114, 117, 122-123, 125-127, 134-135, 143, 144, 145, 146, 150, 156, 162, 165
Space Task Group, 21-25, 28, 159
Sputnik, 11, 19
Stafford, Thomas, 116, 124, 129
Stever, H. Guyford, 44, 104
Stoker, Carol, 25-26
Stategic Defense Initiative, 48, 132
Stuhlinger, Ernst, 18, 19
Sununu, John, 53, 63, 69, 123

T

Technical Study Group, 82-92, 94-95, 97, 100, 102, 104, 105, 111, 117, 128-129, 142, 143, 153, 162, 165
Teller, Edward, 99
Texas A&I University, 114
Thompson, J.R., 58, 62, 130
Tower, John, 48
Townes, Charles, 21
Traxler, Robert, 67, 110-111, 114, 122
Truly, Richard, 2, 53-55, 57-59, 60, 62-64, 67, 69, 70, 73-74, 79-80, 82, 84, 92,

93, 96, 100, 102, 103, 111, 117, 123, 128, 130-131, 139-140, 142, 162-163, 164

U
United States Congress, 4, 19, 21, 23, 28, 30, 37, 39, 46, 47, 49, 51, 52, 54, 55, 66, 67, 69, 74, 75, 77, 78, 82, 93, 95, 97, 98, 103, 105, 109, 110, 111, 113, 114, 117, 118, 119, 120, 121, 122, 124, 125, 126, 128, 133, 139-140, 142, 150-151, 152, 154, 159, 161, 163, 164, 165
United States House of Representatives, see United States Congress
United States Naval Observatory, 7
United States Senate, see United States Congress

V
Viking program, 12-13
von Braun, Wernher, 16-20, 22

W
Walker, Dorothy, 38
Walker, Bob, 120
Wallop, Malcolm, 120
Watson, James, 96
Welch, Steve, 25-26
Welles, Orson, 9
Wells, H.G., 9
Wilford, John Noble, 30
Wilkening, Laurel, 99
Williamson, Ray, 128
Wilson, Roger, 26
Wilson, Pete, 48
Wood, Lowell, 95-96, 99

Y
Yarborough, Ralph, 39
Yeager, Chuck, 28

Z
Zubrin, Robert, 117